COLIN BUCKLE
(formerly Senior Geography Master at Ghana National
College, Cape Coast)

# Landforms in Africa

Longman

Longman Group UK Limited,
Longman House, Burnt Mill, Harlow,
Essex CM20 2JE, England
and Associated Companies throughout the world

First published 1978
Eleventh impression 1992

Set in 10/12 Garamond
Produced by Longman Group (FE) Limited
Printed in Hong Kong

ISBN 0-582-60334-X

*Cover:* Assekrem, black volcanic peaks in the Hoggar Mts, central Sahara.

EP0736

# Contents

Preface

1   Geomorphic processes, rocks and earth          1
    structure
2   The influence of faulting and folding on       12
    landscape
3   Landforms associated with vulcanicity          34
4   Weathering and the movement of material on     61
    slopes
5   River valleys and drainage patterns            79
6   The influence of rocks on landscape            117
7   Plains and inselbergs                          137
8   Desert scenery                                 152
9   Glacial and periglacial scenery                171
10  Coastal landforms                              194
11  Lake basins                                    228

# Acknowledgements

The publishers are grateful to the author and to the following for permission to reproduce photographs in the text:

Aerofilms Ltd., for pages 3 (bottom right), 6, 14, 28, 47, 50, 51, 59, 83, 86, 107, 121, 145, 208, 213; Afrique Photo for pages 7, 35 (top), 137, 153, 154 (top), 155, 156, 164, 167, 169, 197, 203 (top), 212, 234; Michael Crowder for pages 89, 209; East African Railways Corporation, for pages 23, 36, 148; Esso Petroleum Ltd., for pages 3 (bottom left), 63, 154 (bottom); Ghana Information Services, for pages 3 (top), 66 (bottom), 73, 144, 198, 240; Hoa-Qui, for pages 35 (bottom), 55 (top), 77, 78, 110, 115, 118, 134, 230; Institut Geographique National, for pages 30, 45, 48, 119; Kenya Information Services, for pages 41, 84, 173 (bottom), 183, 184, 231; Paul Latham – Uganda, for pages 20, 65, 80, 181, 182, 186, 188, 236; Jacques Martin, for page 129; Ministry of Home Affairs – Rhodesia and S. T. Darke, for page 232; Morocco Tourist Board, for page 75; N.A.S.A., for page 93; Nigeria Information Services, for page 147; Nigeria Magazine, for pages 66 (top), 76, 123, 137, 146, 149, 150, 209; Photographie du Service Photo-Cinema de la Direction de l'Information, for pages 25, 43, 86, 111, 120, 131, 132, 134, 206 (top), 215, 227, 235; Royal Geographical Society, for pages 13, 42, 95, 157, 166, 173 (top); South African Railways, for pages 52, 54, 55 (bottom), 82, 97, 102, 113, 139, 140, 158, 161 (bottom), 163, 190, 200, 205, 221, 239; South African Tourist Corporation – Satour, for pages 31, 100, 221; Tanzania Information Centre, for pages 22 (bottom), 109; H. Tazieff, for page 22 (top); Dr. J. B. Whittow, for page 187; and Afrique Photo for permission to reproduce the cover photograph.

The publishers regret that they have been unable to trace the copyright owners of the following photographs and apologise for any infringement of copyright caused:
Pages 12, 49, 57, 64, 161, 162, 218, 228, 238.

All sketchmaps in the text are aligned north–south, unless otherwise indicated.

Readers unable to obtain specific map sheets from their National Survey Department should contact the following suppliers:

Edward Stanford Ltd,
   12–14 Long Acre, London, W.C.2, U.K.
I.G.N. (Institut Géographique National),
   136 Rue de Grenelle, Paris 7, FRANCE.

# Preface

This book is intended primarily for the student of African geomorphology. Its originality lies in the fact that, with the exception of certain glacial and periglacial features, the examples are drawn entirely from Africa. More than ten years have passed since I first discovered to my amazement that throughout large parts of Africa most physical geography textbooks and much teaching involved the use of European or North American examples. Now, while there is an enormous volume of literature on African geomorphology in various academic journals and research bulletins, these are frequently inaccessible to students or else are published in foreign languages. It is my belief, therefore, that there has long been a need in schools and colleges for a basic text utilising this vast fund of knowledge. I hope that, in part, the present book may satisfy this need.

Throughout the book the aim has been to present geomorphological theory in an abbreviated form, relying on the teacher to provide the circumstances for wider discussion. Where possible each example is supported by a sketchmap or diagram, a photograph, and reference to a survey map sheet. Also I have tried to identify examples in such a way that they can be readily located in school atlases. Names have been, and will no doubt continue to be, a problem. Where alternatives exist the choice is based on frequency of use.

Finally, I should like to acknowledge my thanks to the many people throughout Africa who directly and indirectly helped in the preparation of the book; to the individuals, organisations and agencies who supplied the photographs; to the publishers for their patience and guidance; to J. J. Mensah-Kane, headmaster of Ghana National College; to C. E. Everard, who made innumerable suggestions in the early stages; to Alan Shepherd and Pedro Ankrah for their invaluable assistance; and to my late father for his unfailing support.

Colin Buckle 1976.

# CHAPTER ONE

# Geomorphic processes, rocks and earth structure

Geomorphology is the science that studies the earth's landforms. Landforms are the result of geomorphic processes shaping the rocks of the earth.

## Geomorphic processes

There are two main groups of geomorphic processes: internal, or endogenetic, and external, or exogenetic.

### A. Internal or endogenetic

These originate beneath the earth's surface.

*Earth movements* (Chapter 2)

1 *Faulting* is the breaking or fracturing of rocks. It involves displacement of the rocks on either side of the fracture or fault.

2 *Folding* is the bending of rocks and tends to take place very slowly over many hundreds of thousands of years.

3 *Warping* is the distortion of the earth's surface by uneven upward or downward movement. The movement involved is generally slight, but it may extend over a very wide area. Warping, like folding, is also an infinitely slow process.

*Vulcanicity or vulcanism* (Chapter 3)

Vulcanicity is the movement of molten rock from the earth's interior towards or onto the earth's surface.

### B. External or exogenetic

These occur at the earth's surface (Chapters 4–10).

*Denudation or degradation*

Denudation is the destruction of the landscape by the work of weathering, mass wasting and erosion. Erosion is the wearing away of the land by the natural agents of rivers, waves, tidal currents, glaciers, ice-sheets and wind. The same agents act as a means of transport for the removal of the eroded materials.

*Deposition or aggradation*

Deposition is the complement of denudation. It is the laying down of transported materials by rivers, waves, tidal currents, glaciers, ice-sheets and wind.

## Rocks

The earth consists of 3 main concentric layers: an outer crust, an intermediate mantle, and an inner core. The earth's crust is very thin, being only about 40 km thick compared with the earth's radius of over 6000 km. This crust is in two parts, a lower layer of fairly dense rock and an upper layer of lighter rocks forming the continents.

The rocks of the crust are made up of 8 main chemical elements – oxygen, silicon, iron, aluminium, calcium, potassium, sodium and magnesium. The dense lower layers of rocks, rich in silicon, iron and magnesium, are sometimes described as oceanic since they form the ocean basins.

The lighter overlying rocks are then known as sial or continental. In this context it must be remembered that the true edges of the continent are not the present coastlines, but the lines where the continental shelves slope steeply to the ocean floors.

Rocks, when analysed, are found to be a combination of different minerals, and these minerals are, in turn, formed from one or more of the 8 main elements. Thus granite, for example, is largely a mixture of the minerals quartz, felspar and mica.

| Type | Formed at | Rate of cooling | Crystalline nature | Examples |
|------|-----------|-----------------|--------------------|----------|
| Volcanic | The surface | Fast | Small crystals | Basalt, Rhyolite, Trachyte, Andesite |
| Hypabyssal | Shallow depth | Medium | Medium size crystals | Quartz-Porphyry, Dolerite |
| Plutonic | Great depth | Slow | Large crystals | Granite, Syenite, Gabbro, Diorite |

Fig. 1.1 Types of igneous rocks

All the various rocks found in the continents can be classified, both according to origin and to age.

**Classification of rocks by origin**

There are three main rock types based on origin.

1 *Igneous*: formed by the solidification of molten rock either by intrusion within the earth's crust or extrusion at the surface (Fig. 1.1).
2 *Sedimentary*: formed by the breakdown of pre-existing rocks into sediments, and deposited by water, wind or ice. Over time the sediments become compressed and hardened, and the different grains may be consolidated by a solution of iron, silica or calcium cement (Fig. 1.3).

Sedimentary rocks are formed in layers, known as strata, and the plane between each layer is the bedding plane (Plate 1.1). Earth movements may cause the strata of sedimentary rocks to be tilted out of the horizontal. The angle at which the bedding plane slopes away from the horizontal is known as the dip (Fig. 1.2).

A characteristic feature of some sedimentary rocks is the occurrence of fossils. Fossils are the remains of ancient animal or plant life embedded in the rocks. Few actual remains of plants and

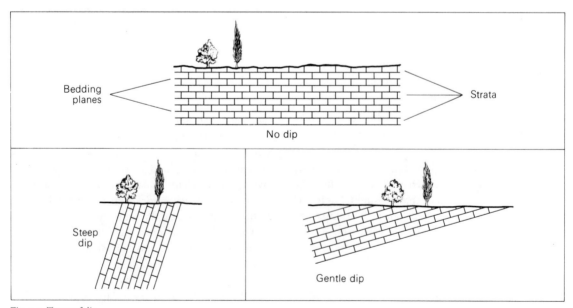

Fig. 1.2 Types of dip

Plate 1.1 Horizontally bedded Voltaian Sandstone at Boti Falls, NE of Koforidua, Ghana

animals are preserved as fossils. Chemical changes usually take place and the original material is substituted by mineral matter, such as silica (Plate 1.2).

3 *Metamorphic:* formed from pre-existing rocks being changed by the influence of great pressure and heat. All types of rock including earlier metamorphic rocks can be changed (Fig. 1.4). The texture and composition of the original rock is altered by pressure associated with earth

| Formation | Nature | Examples |
|---|---|---|
| Mechanical – from the drying out and consolidation of sands, muds, clays and gravels | Argillaceous<br><br>Arenaceous<br><br>Rudaceous | Shale, clay, mudstone<br>Sandstone<br>Tillite, conglomerate |
| Organic – from the remains of once living plant and animal organisms | Calcareous<br><br>Carbonaceous | Limestone<br><br>Coal |
| Chemical – from the precipitation of salts in solution by chemical reactions or evaporation of water | Calcareous<br><br>Saline | Dolomite<br><br>Rock salt |

Fig. 1.3 Types of sedimentary rock

Plate 1.2 Petrified wood in the Libyan Sahara. This fossilised material is formed by the replacement of wood by silica in such a way that the original structure of the wood is preserved

Plate 1.3 Intensely folded Precambrian gneiss on the side of the Azum valley, near Zalingei immediately west of Jebel Marra volcanic upland, Sudan

3

| Original rock | Metamorphic rock |
|---|---|
| Granite becomes | Gneiss |
| Sandstone | Quartzite |
| Clay | Slate |
| Shale | Schist |
| Limestone | Marble |

Fig. 1.4 Types of metamorphic rock

movements, such as folding (Plate 1.3) and heat caused by the intrusion of molten rock. There are 3 kinds of metamorphism:

a) dynamic: pressure force,
b) thermal: heat force,
c) thermo-dynamic: combined heat and pressure force.

When metamorphism results in the rock minerals being recrystallised and formed into separate

| Era | Period | Million years ago | Major geologic events affecting Africa | |
|---|---|---|---|---|
| Quaternary | Recent Pleistocene | | Widespread formation of river terraces and raised beaches. | Glaciation of East African mountains. |
| | | 2 | Deposition of continental sediments in basin areas like the Kalahari. | Crustal down warping of Chad Basin. |
| Cainozoic | Pliocene Miocene Oligocene Eocene | 11 25 40 70 | Main period of volcanic activity in East Africa began. Main period of East African rift faulting began. | Alpine earth movements formed Atlas Mountains. Lava flows in Ethiopia. |
| Mesozoic | Cretaceous Jurassic Triassic | 135 180 | Deposition of marine sediments in many areas, for example South Nigeria. Drakensberg lavas. | Benue Rift formed. Cape Fold Mountains raised up. |
| Palaeozoic | Permian Carboniferous Devonian Silurian Obdovician Cambrian | 225 270 350 400 440 500 600 | Ice age in central and south Africa. Extensive deposition of sediments, for example Voltaian, Bandiagara and Table Mountain sandstones. | Crustal warping causing continental basins. Ice age in parts of North Africa. |
| Precambrian | Upper Middle Lower Azoic | 1000 3000 | Ancient mountain building periods and ancient glaciations. Oldest recorded rocks are 3500 million years, from South Africa. | |

Fig. 1.5 Geological time scale.

layers, the rock is said to be foliated. Fine-grained rocks like schist are usually well foliated.

## Classification of rocks by age

The age of the continental rocks is grouped into four main eras: Precambrian, Palaeozoic (ancient life), Mesozoic (middle life), and Cainozoic (recent life). The last three are also described as Primary, Secondary and Tertiary, and another era, Quaternary, is included to represent approximately the last one to two million years.

Each era is divided into periods or systems. These periods may in turn be subdivided into series and the series may be subdivided into stages or formations. The geological time scale (Fig. 1.5) shows these various time intervals from the earliest known rocks to the youngest formed in the last few thousand years. The geologic events shown on the right of the scale are very approximate and are intended purely as a guide.

The names of the eras and periods shown on the time scale are accepted throughout the world, which means that rocks everywhere can be readily dated on an international basis. The series and formations are not given on the time scale, since these will vary from continent to continent and country to country. In some instances, however, the rocks of a particular period are found to be of sufficient significance to merit a separate name, for example the Karoo System of southern and central Africa. The rocks of this system were formed during a period of time that extended from approximately Carboniferous to Triassic times, and therefore overlapped the Palaeozoic and Mesozoic eras.

The Karoo period may be taken as an example to illustrate the division of systems into series and formations (Fig. 1.6). From this table it will be realised that sometimes a formation is self-explanatory in the type of rock by origin as well as age, while in other instances the variety of rocks involved may be too great, in which case the term 'beds' is often used. Thus, the Molteno Beds are characterised by both coarse-grained sandstones and blue-grey shales. The name Molteno has been derived from the district where the rocks were first recorded. Rock formations are frequently named in

this way, and another example is shown in Fig. 1.7. In some areas the series may be divided solely into upper, middle, and lower formations, according to the positions of the outcrops. Readers are urged to acquaint themselves with the series and formations in their own local area, details of which should be available from their national Geological Survey Department.

| Era | Period | Series | Formation |
|---|---|---|---|
| Mesozoic | Karoo | Stormberg | Drakensberg volcanics<br>Cave sandstone<br>Red beds<br>Molteno beds |
| | | Beaufort | Upper<br>Middle<br>Lower |
| Palaeozoic | | Ecca | Upper<br>Middle<br>Lower |
| | | Dwyka | Upper shales<br>Tillite |

Fig. 1.6 The Karoo System

| Period | Series | Formation |
|---|---|---|
| Devonian | Sekondi | Sekondi sandstone<br>Efia Nkwanta beds<br>Takoradi shales<br>Takoradi sandstone<br>Elmina sandstone<br>Ajua shales |

Fig. 1.7 The Sekondi Series of Ghana

The mixture of mineral compounds that make up rocks is very complex, with the result that any given rock, say granite, will vary from place to place. All granites are largely composed of the minerals quartz, felspar and mica, but from one place to another granite will differ in the quantity and distribution of these minerals, thus Dodoma granite is different from Cape Coast granite.

5

Likewise sandstones will vary in colour, texture and hardness according to factors such as whether they were deposited by water or wind, and what the climatic conditions were at the time.

### Basement rocks

Ancient Precambrian rocks outcrop over a large part of Africa (Fig. 1.8). These rocks are sometimes known collectively as the Basement Complex or Basement Rocks. In addition, much of the area now covered by younger rocks is underlain by basement rocks.

Archean is a term sometimes used to denote the oldest Precambrian rocks of a particular region, but it does not necessarily refer to any special time during the Precambrian.

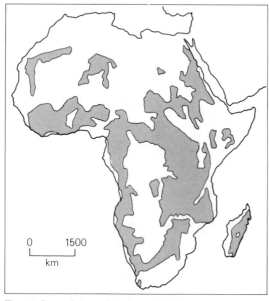

Fig. 1.8 Precambrian rock outcrops in Africa

### Geological time

Geomorphic processes, such as denudation and folding, take place extremely slowly, and it is very important to realise the enormous period of time involved if the origin of Africa's landforms is to be understood.

If the geological history of the earth were compressed into the length of a single year, then the Palaeozoic era would not begin until October and the Quaternary would be shorter than the last hour

of the last day of December. Human history is dated in thousands of years, but geological history is dated in thousands of millions of years.

That the processes of rock formation are continuous is clearly apparent in the deposition of sediments in the Niger delta and the outpourings of molten rock from volcanoes like Nyamlagira in Zaire (Plate 1.4).

Plate 1.4 Part of the vast flow of basaltic lava that reached Lake Kivu after the 1938 eruption of Nyamlagira. The rift scarp is in the background

## Rock jointing

Joints are the small cracks found in nearly all rocks (Plate 4.3). In contrast to faults they involve no displacement. Joints in igneous rocks are due to stress as the rock cools, while in sedimentary rocks they result from hardening and shrinkage during formation. More frequently joints are caused by the tremendous forces exerted on rocks during faulting and other earth movements (page 1).

Joints also develop in rocks due to unloading or the release of pressure. As surface rocks are removed by erosion so the pressure on the underlying rocks is reduced, causing them to split along the lines of greatest pressure-release, that is, parallel to their surface. Pressure release jointing is common in igneous and metamorphic rocks formed under great pressure deep in the crust.

Plate 1.5 Columnar jointing in the basalt cliffs of Cape Manuel, Dakar, Senegal

Certain volcanic igneous rocks, especially basalt, sometimes develop a particular joint pattern in polygonal columns. This columnar jointing results from the rock cooling and contracting about evenly spaced centres causing it to break into a regular polygonal joint pattern. Examples of columnar jointing occur at a large number of places in Africa, for instance Cape Manuel at Dakar, Senegal (Plate 1.5); the Njorowa Gorge south of Lake Naivasha, Kenya; and the Lupata Gorge on the Zambezi, Mozambique (Fig. 5.49).

Rocks that are poorly jointed and with few bedding planes are called massive rocks.

## Water in rocks

When water falls as precipitation from the atmosphere onto the earth it may:
1 be evaporated back into the atmosphere either directly, or indirectly by transpiration through plants;
2 run off into streams and rivers;
3 percolate into the rocks and become ground water.

The amount of evaporation, run-off and percolation depends on the climate, the slope of the land, and

the permeability of the rocks. Permeability describes the capacity of a rock to allow water to pass through it. An impermeable rock does not allow water to pass through.

Permeability is especially influenced by porosity, which refers to the amount of pore space between the rocks' mineral particles. A porous rock is one with pore spaces large enough to hold water. If the pore spaces are also interconnecting, as in many sandstones, water can pass through and the rock is therefore both permeable and porous. Clay is an extremely porous rock, but it is not permeable because the pore spaces are not linked together.

Permeability also depends on whether the rock is pervious, that is, crossed by joints and fissures through which water can pass. Certain rocks, like

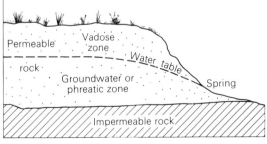

Fig. 1.9 The water table

7

granite and slate, have only a limited pore space, and without joints they are very impermeable.

The depth to which water can percolate is usually limited by an impermeable rock layer. Above this layer the rocks will be saturated to a level known as the water table (Fig. 1.9). The height of the water table in the rock varies with the season. Water circulating through the rocks above the water table is called vadose water. Where the water table meets the land surface springs may occur.

The amount of underground water in an area will depend, therefore, especially on the nature of the rock, that is whether it is permeable or impermeable. A permeable layer that can store water is known as an aquifer. Other factors influencing ground water location include the relief of the area and the rock structure. The level of the water table tends to rise up below a watershed and sink below a valley. Where the rock structure is such that a small impermeable layer occurs within a major permeable rock, it is possible for a localised zone of ground water to accumulate above the impermeable layer and form a perched water table at a higher level than the main water table.

# The structure of continents and ocean basins

The weight of continental rocks (sial) on the underlying oceanic rocks (sima) of the crust (Fig. 1.10) is in a state of balance known as *isostasy*. If the weight of continental rocks is increased in one area due, for example, to a large volcanic eruption or sediment deposition in a river delta, the extra weight disrupts the balance in the area causing a

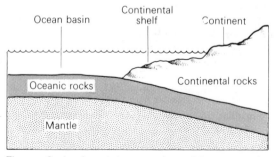

Fig. 1.10 Section through the crust and part of the upper mantle

sinking movement. On the other hand, if the weight of continental rock in an area is decreased by long term denudation the reduced weight causes uplift. These vertical loading and unloading movements are called isostatic adjustments. They take place extremely slowly over thousands of years, but they maintain the balance between oceanic and continental rocks.

## Tectonic plates

Recent geological research shows that the crust is not in one piece but is divided into a number of separate parts called tectonic plates. These plates (Fig. 1.11) are mobile and float on the partially molten rocks of the upper mantle. The lighter rocks of the continents rest on these mobile tectonic plates. Horizontal movement of the plates is extremely slow, in the region of only a few centimetres a year.

It is not known for sure what causes the plates to move, but it is thought that radioactively generated heat in the upper mantle is sufficient to produce a system of slowly turning convection currents which carry along the tectonic plates with the continents as passengers.

The plates are not permanent features, but have changed over geological time in both size and shape. At present there are believed to be 6 main plates and a large number of smaller ones (Fig. 1.11). But the important regions are the boundaries, where plates meet or move apart.

Where plates spread apart lighter elements of hot mantle rock rise to fill the crack creating a new crust. On the sea floor this spreading process produces the mountain ridges that cross the oceans (Fig. 1.12), such as the Atlantic Ridge (Fig. 1.11) which reaches sea level in islands like Ascension. If plates spread apart beneath a continent, the continent will slowly break up and drift apart in smaller sections. The break-up causes faulting and warping of the new land masses.

When plates move together the edge of one is forced to pass beneath the other, and the sinking oceanic rock is destroyed by absorption into the hot mantle. These regions where plates are destroyed are called subduction zones. In the oceans the downward movement creates trenches, for example the Java Trench (Fig. 1.11). Continents are built of

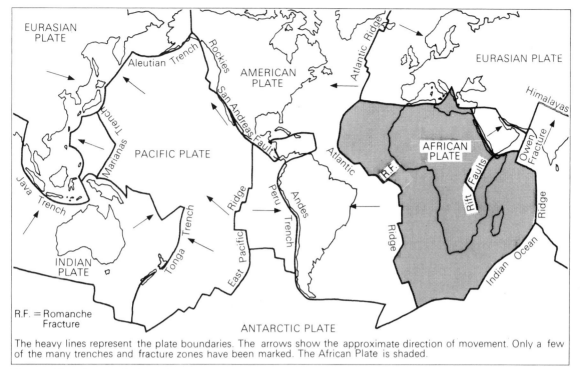

Fig. 1.11 The major tectonic plates of the earth's crust

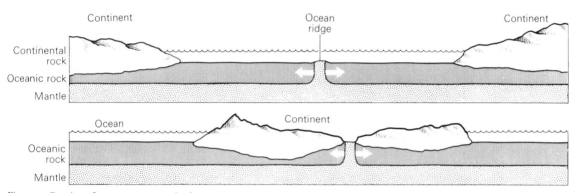

Fig. 1.12 Creation of new crust as tectonic plates move apart

lighter rocks which are not dragged down into the mantle. Instead they collide with each other causing the edges to fold up into mountain ranges, like the Himalayas and the Atlas Mountains (Fig. 1.13).

It is because the continental rocks are light and cannot be dragged down and destroyed in the mantle that they include rocks older than 3500 million years, while the ocean rocks are nearly all younger than 200 million years. For this same reason the continents carry the evidence of past plate movements and former boundaries in the shape of ancient mountain ranges and old fault lines.

A third boundary involves plates moving sideways past each other along transform faults which offset sections of ridges and trenches against one another. At this type of boundary the crust is neither being created nor destroyed. Examples include the Owen Fracture near the entrance to the Gulf of Aden and the Romanche Fracture off West Africa (Fig. 1.11).

9

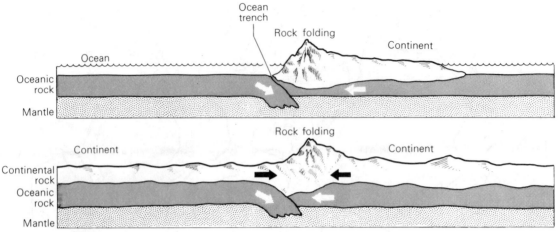

Fig. 1.13 Destruction of crust as tectonic plates move together

All these movements promote earthquakes (page 32) and vulcanicity at and near the plate boundaries, especially in regions where more than two plates meet and they are moving in different directions. The friction of rock surfaces from faulting and other earth movements produces increased heat. This encourages the upward passage of the lighter elements of molten rock from the mantle and causes vulcanicity (Chapter 3).

### Continental drift

The movement of the tectonic plates results in the movement or drift of continents, and there is considerable geological and other evidence to prove that the present continents have drifted apart over millions of years. As continents drift from one latitude to another they experience climatic change. This change is reflected in the rock formations that develop under the different climates. Only continental drift can account for the apparent reversals of climate that have left glacial deposits in the Congo Basin (page 171), coral limestones in north Greenland, and fossilised trees in Antarctica. Further, it is often possible to identify matching rock systems in different continents on either side of the same ocean, and nowhere is this more obvious than between West Africa and Brazil.

But the most conclusive proof of continental drift comes from paleomagnetic dating. When rocks solidify they are magnetised in the direction of magnetic north at that time. By studying the

magnetism of ancient rocks it is possible therefore to determine where on the earth's surface they were originally formed.

Paleomagnetic studies show that the most recent period of continental drift in the earth's history began about 200 million years ago in the early Mesozoic era, when the present continents split off from a single giant continent: Pangaea (a Greek word meaning 'all earth'). The north of Pangaea has been called Laurasia and the south, including what is now Africa, Gondwanaland.

### Geological structure of Africa

The present continent of Africa is thought to have originated in the Cretaceous period as South America, India, Africa, Antarctica and Australia finally broke away from each other (Fig. 9.1). The remnants of this disruption are still to be seen, for as India became detached from Africa and drifted north-east small continental rock masses were left to mark the event. The most significant of these is the block of continental granite 1800 km east of the Kenya coast that forms the Seychelles. The largest island of continental rock off the African coast is, of course, Madagascar.

The break-up of Pangaea involved the formation of great rifts and a warping of the land surface. These upheavals were accompanied by eruptions of molten rock, like the Drakensberg lavas of southern Africa (pages 50–51). Following the break-up, the African tectonic plate collided with the southward moving Eurasian plate. This collision caused ocean sediments trapped between the plates to be folded up during the Cainozoic era into the Atlas Mountains of north Africa (page 29) and the Alpine ranges of southern Europe.

Evidence for continued continental drift is to be seen in the Red Sea, which originated as a rift that began to spread apart about 20 million years ago and is now over 300 km wide. It is an ocean in the making and along its centre is a rift that contains no continental rocks.

In eastern Africa the giant rift valleys indicate the tremendous forces involved in the movements of plates. Here the crust has been upwarped into huge swells. Along the flanks of the swells faults developed following the lines of Precambrian fractures (Chapter 2). As the continent adjusted to the opening of the Red Sea tensional movements along faults caused the rift valleys. Widespread vulcanism is associated with both the Red Sea and the rift valleys (Chapter 3).

Africa also bears the scars of ancient earth movements that occurred before the destruction of Pangaea. Pangaea itself was presumably the result of the assembly of earlier continents that drifted together, and on collision raised up older mountains like the Cape Ranges of South Africa (page 29). Evidence of even older mountains is apparent elsewhere, and the trend of folding can often be identified from the structure, for example the northeast–south-west alignment of Precambrian rock outcrops in West Africa.

Throughout all its long history the continent of Africa has suffered extensive upwarp and depression. The amount of warping was not the same in all areas, and the result has been the formation of a series of basins separated by upwarped higher areas or swells (Fig. 1.14). The effect of long term denudation on this structure means that the swells now consist mostly of Precambrian rocks, while the intervening basins are filled with younger sediments (compare Fig. 1.8 with Fig. 1.14). Some of the basins are quite old, such as the Karoo and Congo, in both of which Palaeozoic sediments outcrop. Due to unequal uplift the Karoo Basin now lies over 1000 m above sea level in contrast to the Taoudene Basin which at its centre is less than 150 m. The diameter of the Taoudene exceeds 1200 km, making

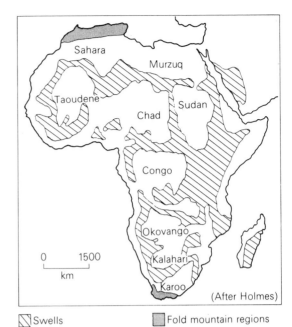

Swells       Fold mountain regions

Fig. 1.14 The major basins and swells of Africa

it one of the largest structural depressions in the world. To the east is the much younger Chad Basin that resulted mainly from Quaternary downwarping.

The present structure of Africa is thus the outcome of processes that have been taking place on a world scale for many millions of years. Structure plays an important part in the landscape of the continent, giving rise to several very individual landforms.

## The geomorphic cycle and climatic change

The landscape is constantly being destroyed by denudation (page 1), especially by stream erosion, and although the processes take place extremely slowly, over long periods of time, a mountainous upland may eventually be worn away to a level plain.

The lowest level to which streams can erode is called the base level of denudation. Most streams flow into the sea and the ultimate base level is therefore sea level. Major streams and lakes act as local base levels for streams flowing into them. A local temporary base level is sometimes caused by a hard band of rock in the river bed below which the river is temporarily unable to erode (page 110).

As the processes of weathering, erosion and deposition take place they give rise to a sequence of events that gradually modifies the landscape. This sequence of changes in the landscape and the landforms that develop at each stage is known as the geomorphic cycle or the cycle of erosion. The cycle as it is known today was first outlined by W. M. Davis who used the humid temperate lands as a model to describe the various stages. The stage of erosion in the cycle refers to the time period a particular landscape has been subject to denudation. The terms 'youth', 'maturity' and 'old age' are often used to illustrate the different stages of development from the beginning to the end of the cycle, for not all landscapes have suffered the same amount of erosion. These stages are not recorded by time in years, but by the amount of work accomplished or to be accomplished.

If all the work of denudation and deposition in an area is accomplished and the landscape reduced to a low-lying level plain then the cycle is complete. However, the completion of a cycle depends on both the climate and the base level remaining unchanged. Any movements of base level or changes in climate will cause interruptions in the progress of a cycle, and for this reason many cycles are only partially completed. Interruptions in the course of a cycle result in the initiation of a new cycle.

Base level movements are due to relative movements in sea level. If sea level falls relative to the land the movement is negative. If sea level rises relative to the land the movement is positive. A negative movement will cause streams to cut down into their valleys to reach the new, lower level. This increased erosion, called rejuvenation, will initiate a new cycle and the whole sequence of events will be repeated, starting again from the beginning. A negative movement may be caused by an actual fall in sea level or an uplift of the land.

Both uplift and depression of the land can be the

Plate 1.6 A raised beach on the South African coast now occupied by the town of Hermanus. Sea level today is some 18 m lower than the former beach with its backing cliffs

result of earth movements, such as folding and warping, or isostatic adjustment (page 8). Changes in sea level are especially related to glacial periods (page 171). Sea level falls due to large amounts of water forming ice sheets. Sea level rises as the ice sheets melt and the water is released. World wide movements in sea level of this nature are known as eustatic changes. The evidence for sea level change is most vividly apparent in the numerous raised beaches (page 217) around the coast of Africa (Plate 1.6).

Climatic change interrupts a cycle because it influences the agents of denudation. The most significant changes in climate of the last two million years occurred during the Pleistocene. In the temperate lands at this time glacial periods alternated with warm interglacials (page 171). During the interglacials the wind belts shifted northwards and tropical Africa experienced pluvial periods when the climate was much wetter than now. These pluvial periods brought savanna vegetation to parts of the Sahara and Kalahari, increasing the number of wild animals and making it possible for men to rear cattle (Plate 1.7). At the same time, higher rainfall raised lake levels and increased the discharge of rivers causing rejuvenation. Rejuvenation also occurred in the glacial periods because sea level was about 100 m lower due to so much water being locked up in the great ice sheets. Throughout Africa, therefore, in a wide variety of landscapes, ranging from river valleys and coasts to high mountains and deserts, there is clear evidence that many landforms were formed under climates different from those operating in the same areas today.

Plate 1.7 Neolithic paintings of cattle on a rock face in the Gilf Kebir Plateau, SW Egypt. In the Sahara the Neolithic age, about 7,000 years ago, experienced a minor pluvial or wet period, making it possible to rear cattle in favoured parts

## References

DEWEY J. F., 'Plate Tectonics', *Scientific American*, Vol. 226, No. 5; 1972.

FURON R., *The Geology of Africa*, 1963.

GILBERT SMITH A. and HALLAM A., 'The Fit of the Southern Continents', *Nature*, Vol. 225, pp. 139–144, 1970.

GROVE A. T., 'The Last 20,000 years in the Tropics', in *Occasional Papers* No. 5 'Geomorphology in a Tropical Environment', British Geomorphological Research Group, 1968.

HURLEY P. M., 'The Confirmation of Continental Drift', *Scientific American*, Vol. 218, No. 4, 1968.

MCKENZIE D. P., DAVIES D. and MOLNAR P., 'Plate Tectonics of the Red Sea and East Africa', *Nature*, Vol. 226, pp. 243–248, 1970.

# CHAPTER TWO

# The influence of faulting and folding on landscape

## Faulting and landscape

Faulting (page 1) has a major effect on landscape for it may cause the uplift, depression or horizontal displacement of large blocks of the crust. In addition, faults rarely occur singly but rather in series.

A fault involving a movement of only a few centimetres can cause the displacement of millions of tons of rock, and although individual displacements are often small the cumulative effect of several movements over long periods of time can lead to the formation of a number of landforms that are essentially structural in nature. Such landforms are the direct result of faulting on landscape.

Landforms also develop from the work of erosion along faults. Faults are lines of weakness in the

Plate 2.1 The steep eastern scarp of the Eastern Rift, north of Nakuru, Kenya

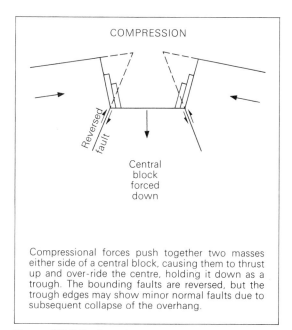

COMPRESSION

Reversed fault

Central
block
forced
down

Compressional forces push together two masses
either side of a central block, causing them to thrust
up and over-ride the centre, holding it down as a
trough. The bounding faults are reversed, but the
trough edges may show minor normal faults due to
subsequent collapse of the overhang.

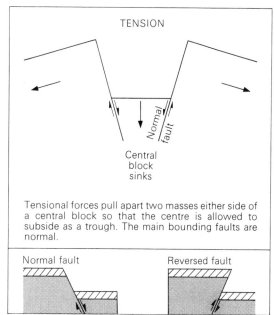

TENSION

Normal fault

Central
block
sinks

Tensional forces pull apart two masses either side of
a central block so that the centre is allowed to
subside as a trough. The main bounding faults are
normal.

Normal fault          Reversed fault

Fig. 2.1  Rift Valley formation theories

crust and as such facilitate erosion, especially as the
rocks along the fault have often been crushed by
the force of the rock masses sliding past each other.
But faulting is most effective in aiding erosion when
rocks of totally different resistance are brought into
contact with one another at the line of a fault, for
erosion naturally attacks the weaker of the rocks
and leaves the harder rock upstanding. Landforms
due to erosion along faultlines are classified as being
the indirect result of faulting on landscape.

The influence of faulting on landscape is thus of a
dual nature: direct and indirect.

## Direct effects of faulting

*Landforms*
Initially, faulted landforms appear as abrupt features, but in time erosion will modify their outlines.
1 *Fault scarp:* an escarpment or scarp is a steep
   slope where the land falls from a higher to a lower
   level. There are different types of scarp according
   to origin, and a fault scarp is a steep slope caused
   by vertical earth movements along a fault (Plate
   2.1). Where movement is only slight or rocks are
   soft erosion may soon destroy the scarp, making

it a short-lived feature. It is important to distin-
guish fault scarps from faultline scarps (pages
25–27), and both from erosion scarps (page 119).
2 *Rift valley:* an elongated trough or depression
   bounded by in-facing fault scarps along more or
   less parallel faults. Rifts vary in width from a few
   hundred metres to several kilometres. Most have
   volcanic activity associated with them. Rift
   origin is not yet fully understood. Several theories have been advanced, most of them involving
   tensional and compressional forces related to
   large scale movements of the crust (Fig. 2.1).
3 *Block mountain* or horst is an upland bordered by
   faults on one or more sides (Fig. 2.2). It stands

Fig. 2.2  Block mountain

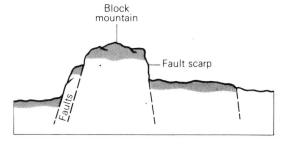

Block
mountain

Fault scarp

Faults

above the surrounding land as a result of being raised or tilted up by earth movements along faults. It may also be due to the relative sinking of the area around a fault-bounded block.

4 *Tilt-block landscape:* a landscape of angular ridges and depressions formed by a series of tilted fault blocks (Fig. 2.3). In western USA, where this type of landscape occurs on a large scale, it is described as 'basin and range country'.

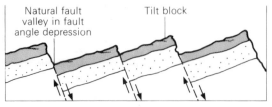

Fig. 2.3 Tilt block landscape

### The East African Rift Valleys

The eastern part of Africa contains a series of rift valleys that are known collectively as the East African Rift Valley System. The whole system is in the form of a number of interconnected troughs in which lie most of the major lakes of East Africa, with the notable exception of Lake Victoria. On

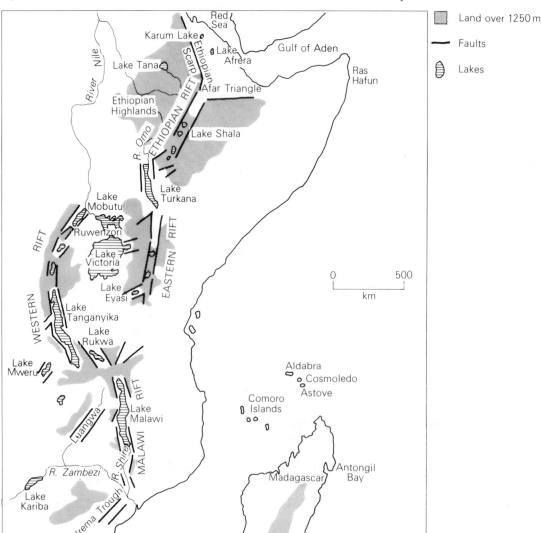

Fig. 2.4 The East African rift system

either side and on the floors of the rifts are hundreds of volcanoes, most of them now extinct (Chapter 3).

The East Africa rifts are the largest and most impressive in the world. Covering a distance of about 4000 km, the system extends from Ethiopia in the north to Mozambique in the south (Fig. 2.4). It can be divided into four main sections:

1 Ethiopian Rift, from the Afar Triangle south to Lake Turkana (Rudolf).
2 Eastern Rift, in Kenya and Tanzania, including the branches in which lie Lake Eyasi and the Kavirondo Gulf of north east Lake Victoria.
3 Western Rift, from Lake Mobutu (Albert) to Lake Tanganyika.
4 Malaŵi Rift, bounding Lake Malaŵi and the Shire Valley.

This section also includes the Urema trough of Mozambique and the Luangwa valley of Zambia. The average width of the rifts is about 50 km and the average height of the bounding scarps is 600 m. But from one section to another there are great variations in both the width and depth of troughs. In Kenya, the Eastern Rift (Fig. 2.5) is up to 100 km wide and the Aberdare Range which forms the east wall rises over 2000 m above the main valley floor (Kenya 1:250,000 sheets SA-36-4, SA-36-8). But south, in Tanzania, rift faulting almost disappears and for several kilometres the scarps barely rise to 100 m.

Such contrasts are even greater in the Western Rift, where Ruwenzori (pages 20–21), a block mountain within the rift, rises over 5000 m and further south the floor of Lake Tanganyika drops to 650 m below sea level.

The nature of the escarpments also differs greatly in the same area from one side of the rift to the other. The Zaire side of the Lake Mobutu rift is nearly twice as high as the Uganda side (Uganda 1:250,000 sheet NA-36-9). In some parts the edge of the rifts are bounded by a single fault, but more frequently there is a system of multiple faults producing a stepped escarpment at the side. Stepped faulting is very marked at the base of the Kedong Scarp in the Eastern Rift, south west of Nairobi (Plate 2.2 and Kenya 1:50,000 sheet 148/3).

Parallel step faults also occur on the west side of the Malaŵi Rift which gives it a more gentle profile than the east. The north-east side of Lake Malaŵi

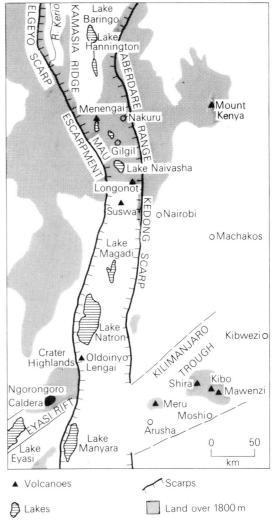

Fig. 2.5 The eastern rift

▲ Volcanoes   Scarps

Lakes   Land over 1800 m

rises sheer from the level of the lake in the spectacular scarp of the Livingstone Mountains to over 2000 m above the lake surface (Tanzania 1:250,000 sheet SC-36-7).

However, although the scarps often rise to impressive heights they, in fact, only represent a fraction of the real depth of the troughs. Over the millions of years since their formation the rifts have been infilled by outpourings of molten rock and great thicknesses of sediments. The depth of infilling is sometimes more than a kilometre. Enormous flows of volcanic rock, such as the Kirikiti and Mbaruk Basalts, have filled the Eastern Rift in

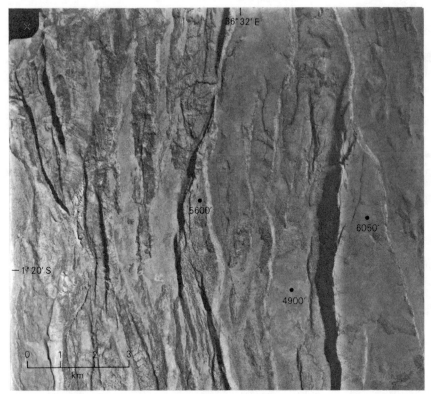

Plate 2.2 Vertical air view of part of the step faulted zone of the Eastern Rift, 15 km west of Ngong, SW of Nairobi, Kenya

Kenya, where the floor is about 1800 m above sea level at Lake Naivasha. Similar volcanic infilling has occurred in the Ethiopian Rift and in the Western Rift north of Lake Kivu (Plate 1.4). Vast quantities of alluvial sediments have also been deposited in the troughs and measurements show that in several places the original floor is below sea level. The depth of Tertiary and Quaternary sediments in the

Lake Mobutu rift exceed 1500 m.

Variations in infilling, together with secondary faults across the rifts, mean that parts of the trough floors are lower than other parts. These relative differences have caused basins in which lakes have formed (page 233). Secondary faulting has also produced fault-guided valleys and minor horsts within the troughs. An example of the latter is the

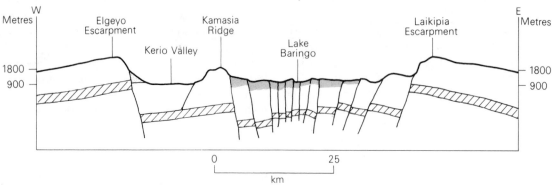

Fig. 2.6 Diagrammatic cross section of the Eastern Rift, Kenya

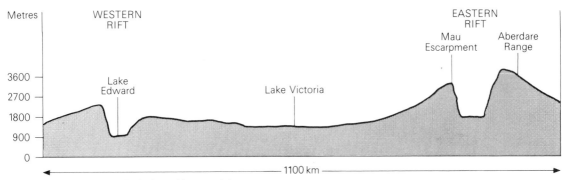

Fig. 2.7 The Lake Victoria Basin formed by crustal downwarp between the western and eastern rifts

Kamasia Ridge (Figs. 2.5 and 2.6) which is separated from the Elgeyo Scarp by the fault-guided Kerio Valley.

The possible origin of the rifts was outlined on page 11, where it was shown that they are thought to be the result of tensional movements along faults on the flanks of huge upwarped swells (Fig. 1.14). They are directly related to crustal fracturing on a continental scale associated with the movement of the tectonic plates (Fig. 1.11). When the crust was uparched fractures developed on the flanks of the swells, and as tension widened the fractures blocks of the crust sank down along parallel faults to produce the rifts. The central part of the main east African swell collapsed forming a shallow basin now occupied by Lake Victoria (Fig. 2.7).

Many of the differences in the rifts are due to variations in the age of each section. While some parts are very old, such as the Shire Valley south of Lake Malawi, others show evidence of recent faulting, for example the steep straight scarps of the Lake Mobutu rift (Plate 2.3). But even apparently recent features may be no more than a rejuvenation of earlier landforms, and in fact many of the rifts are known to coincide with ancient fractures in the basement complex.

Evidence shows that most of the troughs, especially the more recent, are bounded by normal faults pointing to a tensional origin (Fig. 2.1). However, it is thought that older faulting may have been due to compression. The problem with assuming a tensional origin is to understand how the centre block, composed of light continental rocks

(sial), is able to sink down into the normally denser underlying rocks (page 1). Research suggests the existence of a deep layer of light carbonate-rich material in the upper mantle below the region of the rifts. Differences in pressure would cause this light material to rise toward the surface as carbonatite

Plate 2.3 Vertical air view of the fault that borders the east side of Lake Mobutu, Uganda. This section lies between Hoimo and Tonya (Uganda 1:50,000 sheet 38/3)

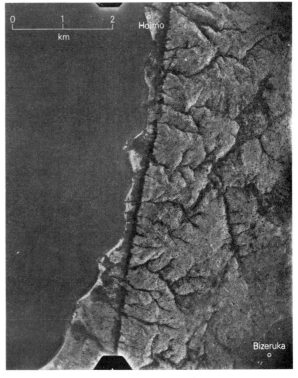

Plate 2.4 Mt Stanley, highest mountain in the Ruwenzori range, viewed from Mt Speke. Part of the Speke Glacier can be seen in the foreground

magma. This, it is believed, would cause localised uplift followed by rock fracturing and the subsidence of a long narrow valley along normal faults. The existence of the carbonatite magma below would allow the centre block (Fig. 2.1) to subside without difficulty.

Much of the volcanic rock ejected from the volcanoes along the rifts includes elements of carbonatite magma. In some, notably Oldoinyo Lengai (page 44), carbonatite plays a dominant role.

*The Benue Rift, Nigeria*
The Benue Valley is not bordered by any visible faults, but the folded sedimentary Cretaceous rocks within it contrast sharply with the Precambrian crystalline rocks on either side. The distance across the valley between the erosion scarps (page 119) of the basement rocks is up to 100 km, while the thickness of the enclosed sedimentary rocks is over 4000 m. The valley may, therefore, be an ancient rift of Cretaceous age, formed as South America and

Africa drifted apart from each other in the disruption of Gondwanaland (page 10). The opening of the rift would lead to flooding by the sea and the consequent deposition of the sediments. Later, when two continents finally broke apart, the walls of the rift may have closed slightly, causing the folding of the sandstones and other sediments.

*Ruwenzori*
The Ruwenzori Range, (Uganda 1:250,000 sheet NA-36-13) is a huge block mountain of Precambrian rocks lying at the junction of two branches of the Western Rift, between Lakes Mobutu and Amin. Ruwenzori rises to over 5000 m and is the highest non-volcanic mountain in Africa (Plate 2.4).

The uplift of this giant structure, which is about 100 km long and up to 45 km wide (Fig. 2.8) occurred mainly in the Pliocene and Quaternary, and its major peaks stand more than 4000 m above the floor of the Semliki Valley to the west.

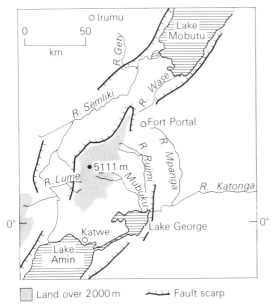

Fig. 2.8 The Ruwenzori Range

highest point, 5111 m. All six mountains are glaciated above 4200 m (Chapter 9).

Other examples of block mountains are Danakil Alps, east of the Danakil Depression in the Afar Triangle, Ethiopia (Fig. 2.4); Usambara Mountains, north-east Tanzania; and Karas Mountains, south Namibia (SW Africa) (Fig. 8.2).

*Tilt block landscape, Ethiopia*

In the Afar Triangle (Fig. 2.4) large blocks of the crust have been tilted and down-faulted on a giant scale. This is especially marked in the region of the lava fields extending south of Lake Afrera, where it produces a rugged, tilt block landscape covering an area of several kilometres. The faulting is roughly aligned NNW-SSE and subsidence has left the summits of the blocks at a lower level than the surrounding plateau (Plate 2.5).

## Fault scarps in Africa

The uplift was greater on the western and northern edges, causing slopes to be much steeper on the Zaire side than on the Uganda side (Fig. 2.9).

Ruwenzori is composed largely of ancient schists and gneisses surrounding a core of granites, diorites and other igneous rocks. It is the igneous core that forms the six chief mountains of Ngaliema (Stanley), Baker, Speke, Emin, Gessi and Luigi di Savoia. They are separated from each other by deep glaciated valleys and high passes. Mount Ngaliema (Stanley) is the highest and has Margherita, the

The East African rifts provide numerous examples of fault scarps in the bounding walls of the troughs, such as the Elgeyo Escarpment (Plate 2.5 and Fig. 2.6), and the Chunya Scarp, southern Tanzania (Plate 2.6).

Sometimes the slopes are very steep and show little evidence of erosion, for example the Manyara Scarp, Tanzania, and the Butiaba Scarp, Uganda (Plate 2.3 and Uganda 1:50,000 sheet 38/3). Elsewhere, scarps are in various stages of dissection. Stream erosion has cut the scarp of the Livingstone

Fig. 2.9 Cross section of Ruwenzori

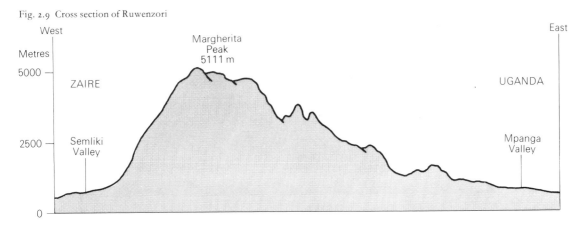

Plate 2.5  Rugged tilt-block landscape lying about 100 km SE of Lake Afrera, in the Afar Triangle of Ethiopia. In places the blocks are whitened by salt

Plate 2.6  The Chunya fault scarp bounding the Western Rift 40 km north of Mbeya, between Lake Tanganyika and Lake Malaŵi

Mountains, on the north-east of Lake Malaŵi, into a series of triangular facets.

Fault scarps are also prominent features along the edges of Africa's block mountains. The faulting on the west of Ruwenzori is fairly recent and erosion has only just begun, but in the north the scarps have been cut back into rounded spurs.

One of the highest and steepest fault scarps in the world is the Ethiopian Scarp which borders the Ethiopian Highlands on the east (Fig. 2.4). In places there is a sheer drop of 1500 m to the Danakil Lowlands at the foot.

Fault scarps also occur in other parts of Africa, notably in the faulted zone of south Cameroon (Fig. 3.5). Here the Bamenda Scarp, above which stands the town of Bamenda, can be traced for nearly 50 km (Cameroon 1:200,000 sheet NB-32-11).

## Drainage

In some regions faulting and warping (page 1) have a significant influence on rivers and drainage patterns.

1 Vertical faulting across a river valley may cause a waterfall (Fig. 2.10A).
2 Tear faulting, a horizontal movement (Fig. 2.10B) across a river will cause the river to be offset at the point it crosses the fault.

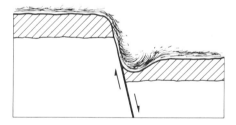

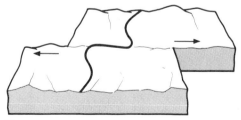

Fig. 2.10

(A) Waterfall caused by faulting across a river valley

(B) River offset by faulting across its valley

3 Rift faulting that forms an enclosed basin may cause a lake if rivers flow into the basin (page 233).
4 Rivers may follow straight, natural fault valleys, especially in areas of tilt-block faulting (Fig. 2.3) and where differential movements have raised some parts while lowering others.
5 Crustal tilting or upwarping across a river valley will gradually force the river to reverse direction and run backwards if it is unable to maintain its original flow.

### Kalambo Falls, Zambia and Murchison Cataracts, Malaŵi

Kalambo Falls (Plate 2.7) have a vertical drop of 212 m. The falls are near the south-east end of Lake Tanganyika, 30 km north of Mbala and are a result of the river dropping over the scarp into the

Plate 2.7 Kalambo Falls, one of the highest waterfalls in Africa, drops from the plateau of central Africa into the rift valley

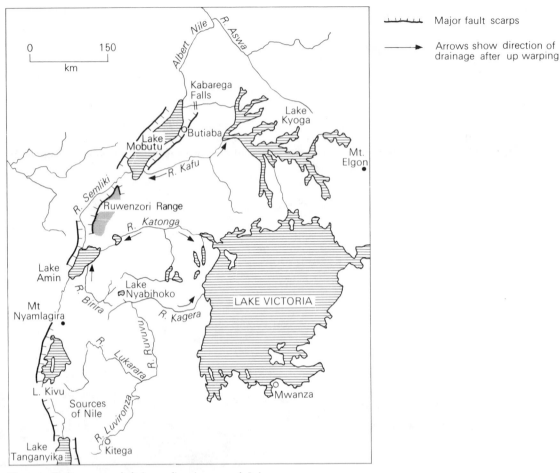

Fig. 2.11 Fault scarps and drainage diversion around Lake Victoria

Western Rift (Zambia 1:50,000 sheet 0831 C1). Above the falls the Kalambo meanders in wide flat valley along the Tanzania-Zambia border. Below the falls it follows a narrow, 5 km long, deep gorge, cut along the line of a fracture in the rocks, before crossing the shore zone into Lake Tanganyika.

From the south end of Lake Malaŵi the Shire River drains south to the Zambesi. In the middle section of the Shire, some 150 to 250 km south of Lake Malaŵi, the river drops 375 m in about 80 km by a series of five rapids known as Murchison Cataracts. The rapids are due to geologically recent movements in the parallel faults which cross the valley at this point.

*Reversed drainage in Uganda*

The most recent period of faulting in the Western Rift of Uganda was followed by slight uplift of the crust to the east. This crustal warping caused streams previously flowing west into the rift valley, like the Kafu, Katonga and Kagera, to reverse direction and flow east. Lake Kyoga and its neighbours north of Lake Victoria are a direct result of this action (Fig. 2.11). Much of Kyoga is a vast papyrus swamp less than 7 m deep, and its peculiar outline is due to the Kafu River being forced to flow back up its own valley and tributaries.

24

## Indirect effects of faulting

1 *Faultline scarp:* a steep slope developed along an ancient faultline by differential erosion of hard and soft rocks on either side of the fault. It is not caused by any recent movement of the fault.

A scarp facing the same direction as an original fault scarp is known as a normal or consequent faultline scarp (Fig. 2.12).

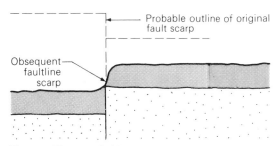

Fig. 2.13  Obsequent faultline scarp

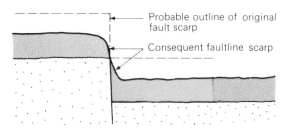

Fig. 2.12  Consequent faultline scarp

A scarp facing the opposite direction to the original fault scarp is called a reversed or obsequent faultline scarp

2 *Fault-guided valley:* faults cause rocks to be shattered and crushed, which means that such rocks are more easily eroded than those further from the fault. Rivers are frequently guided by these zones of broken rock and carve long straight valleys, often intersecting at right angles. In extreme cases this can result in a rectangular drainage pattern (page 105) covering a very wide area. Faulted zones of shattered rock can also play a major part in the development of river capture (pages 101–103).

In coastal regions the alignment of cliffs is sometimes related to wave erosion along faultlines.

*Faultline scarps in Madagascar, South Africa, and Zambia*

The Isalo Massif is a large sandstone upland in south Madagascar, west of Ranohira, between the

Plate 2.8  The east-facing fault-line scarp of the Isalo Massif, near Ranohira, Madagascar

Mangoky and Onihaly Rivers (Madagascar 1:100,000 sheets I-55, I-56). Parts of the massif have been severely dissected and in places rivers have carved deep gorges. But north of Ranohira (Fig. 2.14), dissection is less marked and the massif is bordered on the east by a steep reversed faultline scarp (Plate 2.8). The height of the scarp increases northwards. Near Ranohira it is 150 m, but north of Morarano it is over 500 m. The original downfaulted Isalo sandstone now forms the upland (Fig. 2.15), while the weaker Sakamena clays and shales have been eroded to make the lowland followed by the Menamaty River. The very steep nature of the escarpment is mainly due to the fact that it is being worn back by the splitting off of huge vertical slabs of sandstone. These break up and form vast talus slopes (page 62) of broken rock at the scarp foot, sometimes totally covering the Sakamena rocks.

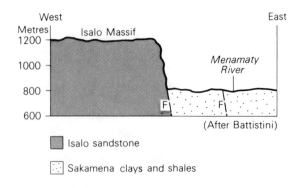

Fig. 2.15 Section of Isalo faultline scarp

Table Mountain, in Natal, South Africa, lies 20 km east of Pietermaritzburg. (Not to be confused with Table Mountain near Cape Town.) It is a flat-topped hill rising to 945 m and lies between the incised meanders (page 99) of the Mgeni and Msunduze Rivers (South Africa: 1:50,000 sheet 2930 DA). The 3 km long steep western edge is a consequent faultline scarp, evident from the fact that the same Table Mountain sandstone that forms the upland has been faulted down and outcrops at the base near the Cacaca Stream, a tributary of the Msunduze. The former cover of Dwyka tillite rock has largely been eroded, except for a small capping on top of the mountain (Fig. 2.16).

In Zambia the Muchinga Escarpment (Zambia 1:500,000 sheets 6 and 9), on the north side of the

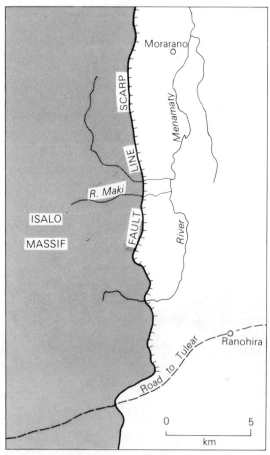

Fig. 2.14 Isalo faultline scarp Madagascar

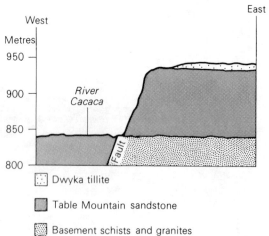

Fig. 2.16 Section of Table Mountain faultline scarp, South Africa

Luangwa Valley (Fig. 2.4) is, in effect, a faultline scarp. The Luangwa trough is a rift valley of Jurassic age that was later filled by sediments. Erosion has slowly removed these sediments so that the original fault scarps, especially the Muchinga, have been resurrected and now form prominent features several hundred metres high, that can be classed as faultline scarps.

Other examples of faultline scarps are: Kilosa-Msolwa Scarp, south-west of Morogoro, Tanzania and Mutito Scarp, on the east side of the Kitui Hills, Kenya (about 150 km east of Nairobi).

### The influence of faulting on the Madagascar coast

In north-east Madagascar the straight sides of Antongil Bay (Fig. 2.4) where granites reach the sea as cliffs over 150 m high, are related to faultlines which border the bay. In fact, much of the east Madagascar coast, especially between Fort Dauphin and Foulpointe, appears to be fault aligned approximately NNE-SSW. This line would seem to be a southern extension of the Owen Fracture (Fig. 1.11), that crosses the floor of the Indian Ocean.

### Fault-guided valleys in Africa

On many rivers there are sections where the valley follows the shattered zone along the line of a fault. East of Kindia, Guinea, the Santa River, a tributary of the Great Scarcies, is guided by a faultline where Palaeozoic sandstones have been downfaulted against basement rocks (Fig. 2.17) (Guinea 1:50,000 NC-28-XVIII-1a).

In Sudan the course of the Nile in the huge bend between the third and the sixth cataracts is related to

the NW–SE and NE–SW rectilinear pattern of faultlines (Fig. 2.18).

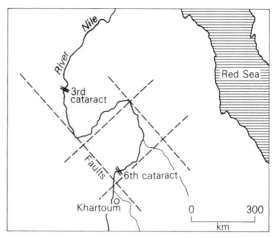

Fig. 2.18 The influence of faulting on the course of the River Nile in Sudan

On the Zambesi the Batoka Gorge, below Victoria Falls (Fig. 5.45) is partly the result of rock fracturing by a number of intersecting faults (Plate 2.9). (Zambia 1:5000 sheets LL 7616 and LL 7618).

Other examples of fault-guided valleys are: the Aswa River, north Uganda, which follows a north-west trending tear fault (Fig. 2.11); the Imo River downstream of Umuahia, south-east Nigeria (page 104) and the Klein Berg River, near Tulbagh, South Africa (page 102).

## Folding and landscape

The folding of rocks (page 1) has produced a variety of geological structures. In some areas the folding was gentle, resulting in simple upfolds and downfolds called anticlines and synclines, but elsewhere it was of such intensity that rocks were folded over, faulted and crushed, so that old rocks are now found on top of younger rocks.

Rock folding is due to the tremendous compressional forces that develop as the tectonic plates of the crust (page 8) move towards each other. Sediments deposited on the floors of seas between two colliding continents are very slowly folded up

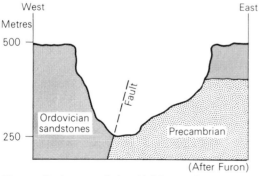

Fig. 2.17 Section across fault-guided Santa Valley near Kindia, Guinea

27

Plate 2.9 Victoria Falls on the Zambesi showing the first four fault-guided arms of the Batoka Gorge below the falls

until they rise above sea level. The process continues and eventually the rocks become so intensely folded and over-folded that they are pushed up to form high mountains, crushing formerly horizontal strata and causing it to dip vertically into the air (Plate 2.11).

The processes of erosion and folding go on simultaneously, and as rocks are folded up into mountains so erosion gradually wears them down. However, at certain periods in geological history folding has clearly been at a faster rate than erosion, with the result that over millions of years mountains like the Atlas and the Cape Ranges are folded and raised up.

The movements involved are infinitely slow and very powerful rivers are sometimes able to continue flowing across the newly-forming mountains, because they cut down into their valleys at a faster rate than the mountains are folded up (page 106).

## Fold mountains

The most recent mountain building period was the Alpine which began in the Tertiary and formed huge mountain chains like the Himalayas and the Atlas (Fig. 1.5). These Alpine ranges are called the Young Fold Mountains. An earlier period was the Hercynian in the Mesozoic era which built mountains like the Southern Appalachians and the Cape Ranges. Much earlier in the earth's history, during the Precambrian, even older fold mountains were raised up. In parts of Africa remnants of these ancient mountain chains are still found, though greatly reduced in height by erosion, for example the Akwapim Hills of south Ghana.

High mountains are the most obvious effect of folding on landscape, but they are not the only effect. On the borders of fold zones, and in areas far removed from major folding, lesser earth movements may produce a series of gentle anticlines and synclines. This is the landscape pattern, now scarred by erosion, to be found in parts of the wide Benue Rift and in south-east Nigeria (pages 122–123).

The relationship between the folding and the topography of an area tends to depend on the intensity of the folding and the resistance of the rocks. In many parts of the main mountain ranges the folding was so great and so complex that there is no direct relation between the folding and the relief. However, in regions where folding was simple and the rocks are relatively resistant a topography may develop where anticlines form uplands and synclines form valleys. In time, as erosion wears away a landscape, it is possible for the relief to be completely inverted (page 106) so that upfolds form lowlands and downfolds form uplands, known respectively as anticlinal valleys (Fig. 5.38) and synclinal uplands (Fig. 2.19).

Folding, therefore, can produce very complicated regions of relief, and fold mountain ranges generally comprise several mountain chains as well as a number of more level areas called intermontane plateaus.

*Atlas Mountains*
The Atlas are the highest fold mountains in Africa, rising to more than 4000 m. They extend from Morocco to Tunisia, a distance of about 2250 km (Fig. 2.20). The folded rocks are mainly Cretaceous and Tertiary sediments, such as sandstones, limestones, and clays, but they also include Precambrian igneous rocks. The most important period of folding was in the mid-Tertiary.

The mountains are not one continuous upland but a series of SW–NE trending chains which enclose several intermontane plateaus and basins. The highest mountains are in Morocco, especially the High Atlas (Plate 2.10). Further east, the Tell and Saharan Atlas in Algeria are much lower (1500 m) and are really a series of small uplands separated from each other by intervening valleys and plateaus. Between the Tell and Saharan Atlas is the Plateau of the Chotts, a huge intermontane plateau with an average height of 1000 m. It is an area of rigid basement rocks that shows only slight evidence of folding, since it was able to resist the main mountain-building forces. The chotts are shallow salt lakes, the largest of which is Chergui (page 233).

Not all the Atlas ranges are the same age, the Rif being much older than the Tell. The Rif also suffered the most intense folding, so that rocks were greatly over-folded and faulted (Fig. 2.21).

*The Cape Ranges*
In the extreme south of Africa, between the Great Escarpment and the sea, Palaeozoic rocks have been folded up to form the Cape Ranges (Plate 2.11). Since their formation in Triassic times, erosion has greatly reduced the height of these mountains, but

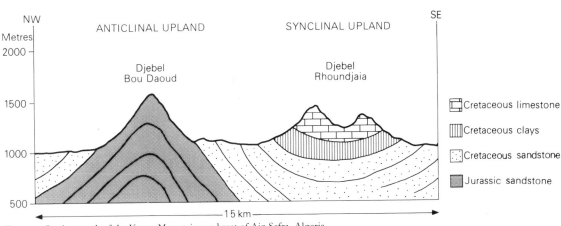

Fig. 2.19 Section north of the Ksour Mountains and east of Ain Sefra, Algeria

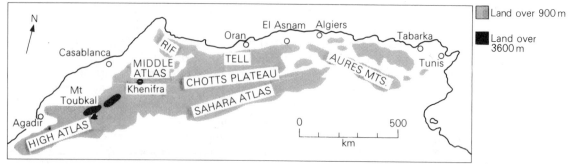

Fig. 2.20 Atlas Mountains

Plate 2.10 Toubkal (4,165 m), one of the main peaks in the High Atlas. Folding has raised the basement rocks and Mesozoic sediments to form a range of snow-capped mountains

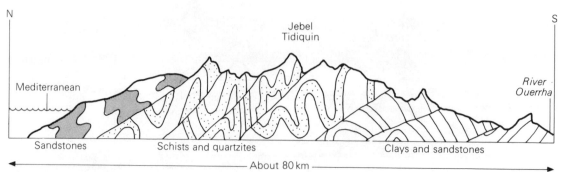

Fig. 2.21 Section through the heavily faulted and folded Rif Atlas Mountains

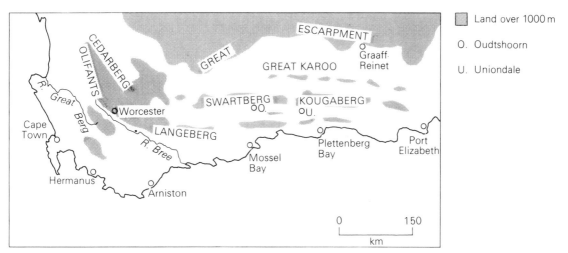

Fig. 2.22 Cape ranges, South Africa

they still rise to over 2000 m. The folds are aligned in two directions which meet in the area of the Hex River Mountains, north of Worcester. There is a north-south trend consisting mainly of the Olifants Mountains and the Cedarberg, and an east-west trend which includes the Swartberg, the Langeberg and the Kougaberg (Fig. 2.22).

The Cape Ranges have been extensively faulted as well as folded, especially along their southern edges. However, the present pattern of relief is a result of both the structural influence of folding and faulting and also the differential erosion of the rocks. The main ranges are largely anticlines of resistant Table Mountain Sandstone and the valleys are synclines from which soft Bokkeveld Shales have been eroded, for example the Hex River Valley bordered on the north by the Hex River Mountains (Fig. 5.30) and on the south by the Kwadouw Range. But not all the ranges are anticlinal and one fine example of a synclinal upland is the famous Table Mountain at Cape Town, now separated by denudation from the main mass of the Cape Ranges. Likewise, as a result of erosion, some valleys are carved along anticlines. The Tulbagh Valley (Fig. 5.30) is an anticlinal valley.

## Domes

Domes also are due to vertical uplift in the crust. The most common causes of doming are the intrusion of molten rock (page 52) and the up-welling of materials lighter than those around them.

Plate 2.11 Steeply folded rocks of the Table Mountain series in the Swartberg Pass of the Cape Mountains, South Africa. The folding was so intense that the rocks now rise vertically into the air

31

Structures of the latter type are called diapirs and they are commonly built of rock salt that rises to the surface from underground deposits. When they uplift the overlying rocks salt diapirs are called salt domes. In Ethiopia a large salt dome lies south-east of Karum Lake (Lake Asale), near Gada Ale volcano, in the Afar Triangle (Fig. 2.4).

## Earthquakes

An earthquake is a shaking of the ground due to a sudden and rapid displacement of rocks beneath the earth's surface. The shaking is caused by a series of shock waves that are sent out from the focus, or point of origin, of the quake.

The majority of earthquakes are associated with faults in the crust and the most severe take place at the boundaries of the main tectonic plates (Fig. 1.11). Much of Africa lies outside the world's major earthquake areas (Fig. 2.23), the notable exceptions being the East African Rift System and north-west Africa.

Evidence of even the slightest earth tremor is obtained by means of an instrument called a seismograph which records the shock waves emitted by the quake. Several thousand earthquakes are recorded every year, but only a few are noted for their destructive effects.

The size of an earthquake is usually measured according to magnitude on the Richter scale and according to intensity on the Mercalli Scale. The Richter scale is based on instrument recordings and ranges from 0 to over 8.0. The Mercalli scale is based on observed effects of damage and changes in the earth's surface, and ranges from 1 (detectable by seismograph) to 12 (catastrophic).

*Earthquakes in Africa*
Only three earthquakes in Africa this century have recorded a magnitude of 7·0 and above on the Richter scale. They were:

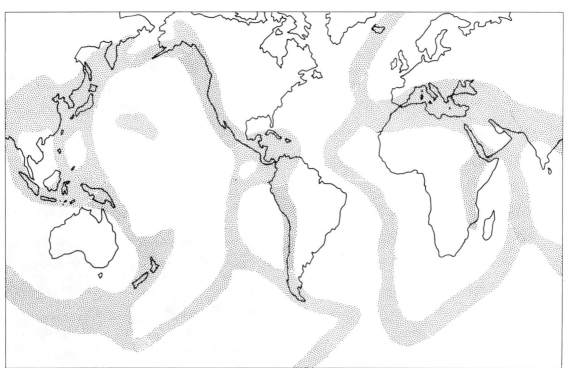

Note: It is largely from the distribution of earthquakes that the boundaries of the tectonic plates (described in chapter one) have been determined.

Fig. 2.23 World distribution of major earthquakes

| Tanzania | 13 Dec 1910 | 7·3 | 20 km NW of Kasanga on the shore of Lake Tanganyika |
| Kenya | 6 Jan 1928 | 7·0 | Laikipia Scarp, east of Lake Hannington in the Eastern Rift |
| Libya | 19 Apr 1935 | 7·1 | Al Qaddahiya, 100 km south of Misratah (Misurata) |

Recent African earthquakes of intermediate magnitude are listed below. The concentration in the rifts and in north-west Africa is apparent. However, beside these two main regions there are also several areas of minor seismic activity, ranging from South Africa to Ghana.

Most of the quakes of magnitude above 5·0 (Richter) have produced minor changes in the landscape, such as landslides and miniature fault scarps. Earthquakes that result in severe damage and great loss of life are fortunately few in Africa. There have been two during the present century: Agadir and El Asnam (Orleansville).

| *Recent African Earthquakes of magnitude exceeding 5·0 (Richter)* | | | |
|---|---|---|---|
| Algeria | 9 Sept 1954 | 6·7 | El Asnam (Orleansville) |
| Morocco | 29 Feb 1960 | 5·7 | Agadir |
| Ethiopia | 14 Jul 1960 | 6·3 | near Lake Shala, 200 km SW of Addis Ababa |
| Ethiopia | 2 Jun 1961 | 6·4 | Karakore, 200 km NE of Addis Ababa |
| Libya | 21 Feb 1963 | 5·3 | Al Marj, north Cyrenaica |
| Tanzania | 7 May 1964 | 6·4 | 60 km SW of Lake Manyara |
| Algeria | 1 Jan 1965 | 5·5 | M'Sila, 185 km SE of Algiers |
| Uganda | 20 Mar 1966 | 6·8 | near Kichwamba |
| Sudan | 9 Oct 1966 | 5·7 | south of El Obeid |

At Agadir the focus was of shallow depth and sited directly beneath the town. Nearly all of Agadir was destroyed, 12,000 people died and at least the same number were injured. Additional damage resulted from a tsunami, or giant sea wave, caused by the earthquake, which swept inland for one third of a kilometre.

The El Asnam earthquake destroyed an area of radius 40 km (Fig. 2.24). It caused landslides and opened fissures in the ground more than 3 m deep.

After Agadir, *The Times* of London reported: 'Evidence of the tremendous upheaval off the coast has come to light from preliminary naval soundings, which show in one case a depth of only 45 feet where previously the water was charted at 1200 feet.

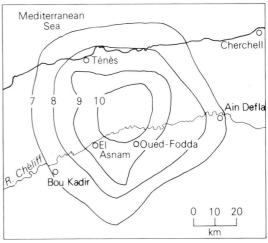

Fig. 2.24 Isoseismal lines on the Mercalli Scale for the Algeria quake (El Asnam) 1954

In a position approximately 9 miles from the shore, soundings show a depth of 1200 feet instead of 4500 feet.'

## References

BATTISTINI R. and DOUMENGE F., 'La Morphologie de l'Escarpement de l'Isalo et son revers dans la région de Ranohira', *Revue de Géographie Madagascar*, No. 8, pp. 67–92, 1966.

DOORNKAMP J. C. and TEMPLE P. H., 'Surface Drainage and Tectonic Instability in part of southern Uganda', *Geographical Journal*, Vol. 132, pp. 238–252, 1966.

FREUND R., 'Rift Valleys', in *The World Rift System, Geological Survey of Canada*, Paper 66–14, 1965.

GIRDLER R. W., 'Drifting and Rifting in Africa', *Nature*, Vol. 217, pp. 1102–1106, 1968.

HOLMES A., *Principles of Physical Geology*, Chapter 29, 1965.

KING L. C., 'Speculations on the outline and mode of disruption of Gondwanaland', *Geological Magazine*, Vol. 87, pp. 353–359, 1950.

ROSE C., 'The Kalambo River and Falls: The Geology of the Kalambo Gorge', *Geographical Journal*, Vol. 74, 1929.

TAZIEFF H., 'The Afar Triangle', *Scientific American*, Vol. 222, No. 2, 1970.

# CHAPTER THREE

# Landforms associated with vulcanicity

Vulcanicity or vulcanism involves the processes through which gases and molten rock are intruded into the earth's crust or extruded onto the earth's surface. It includes all processes resulting in the formation of volcanoes, lava plateaus, and intrusive igneous features.

The molten rock, known as magma, originates in the upper plastic layer of the earth's mantle (Fig. 1.10). The mantle rock is very dense in nature compared to the crustal rocks, but due to high temperatures it is near to melting point. Additional heat, sufficient to take the material to melting point, is generated by friction along rock surfaces at the boundaries of the tectonic plates that form the earth's crust (page 8). The friction is largely the result of faulting and other crustal movements. In fact, volcanic activity is nearly always associated with major fault zones, though not all fault zones give rise to vulcanicity. With increased temperatures the lighter elements of the dense mantle melt and begin to rise towards the surface as magma. The molten rock rises due to the much lower pressure at the surface and as it does so it forces its way along fissures in the continental rocks created by the explosive activity of its own escaping gases.

When magma erupts at the surface and loses its gases it is known as lava. It is the escaping gases, especially steam, which, because they expand rapidly due to the lower pressure, cause many eruptions to be very explosive. Lavas vary considerably in their chemical composition, particularly in the silica content, and thus in their degree of mobility. The main types of lava are shown in Fig. 3.1.

Intermediate and acid lavas, because of their viscosity, may block the fissures and are usually associated with explosive types of eruption. When an eruption takes place explosively the magma is often ejected in the form of ashes, cinders and blocks, of which the finest is ash. These fragmental materials are referred to as pyroclasts. Eruptions that take place quietly with almost no explosive activity are described as effusive.

| Type | % Silica | Degree of mobility | Example of rock type |
|------|----------|--------------------|-----------------------|
| Acid | More than 66 | Extremely viscous and immobile. Solidifies rapidly at high temperatures of over 850°C | Rhyolite |
| Intermediate | 52–66 | Fairly viscous. Unable to flow far before solidifying | Trachyte |
| Basic | 45–52 | Very fluid and mobile. Able to flow long distances before solidifying | Basalt |

Fig. 3.1 Types of lava

However, not all magma reaches the surface as lava. Some cools and solidifies within the crust in the form of intrusive igneous rocks.

The result of these various processes of vulcanicity is the development of two major types of vulcanic feature: extrusive landforms and intrusive landforms.

## Extrusive landforms

### A. Volcanoes

A volcano is a hill or mountain formed by the eruption of molten rock from a central opening or vent in the crust (Fig. 3.2). The material erupted builds up around the vent, generally in the form of a cone or dome with a funnel-like depression called a crater at the summit (Plate 3.1). As long as there is a supply of magma a volcano will continue to grow to a height to which differences in pressure can force the erupted materials. If the pressure becomes insufficient for the magma to reach the main crater the molten rock may force its way to the surface through other fissures and build parasitic or adventive cones on the slopes of the main volcano (Plate 3.2). Parasitic cones may also develop if the chief vent becomes blocked. When magma solidifies along fissures it forms dykes.

The size and shape of a volcano is mainly determined by the nature of the material erupted and the mode of eruption. Volcanoes vary in size from small cones only a few metres high to vast mountains like Kilimanjaro (Plate 3.8). They are very complex structures and any classification is bound to be arbitrary. Over a period of time a volcano may change its mode of eruption and thus its general characteristics.

The major types of volcano are illustrated in Fig. 3.3, with a summary of their main features. In addition, there are residual volcanic landforms that

Plate 3.1 Kibo, the summit of Kilimanjaro, with its pit-like crater

Plate 3.2 The upper part of the active volcano of Piton de la Fournaise, Reunion. Note the large sunken crater and numerous parasitic cones associated with fissure eruptions on the flanks

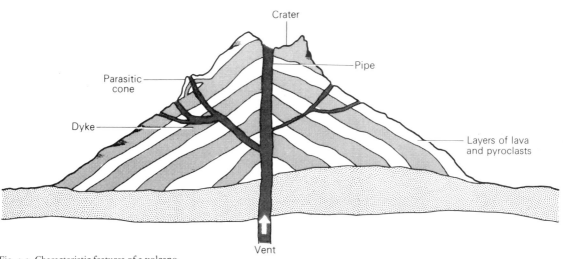

Fig. 3.2 Characteristic features of a volcano

result from the destruction of a volcano. These are shown in Fig. 3.4.

### Active, dormant and extinct volcanoes

According to their present state of activity volcanoes are described as being active, dormant or extinct. Extinct means the volcano now shows no sign of further eruption and much of the original structure may have been destroyed by denudation. Dormant describes a volcano that has not been known to erupt and yet is not thought to be extinct. Some volcanoes show only limited activity in the form of fumaroles, for example Kilimanjaro. Fumaroles are small subsidiary vents from which issue mainly gases. Broadly speaking, active describes a volcano that has erupted or is thought to have erupted during the last 500 years.

Most volcanoes in Africa are now extinct or dormant, and at present there are less than 50 which can be called active. The important ones are listed below.

| *Volcanoes classified as active in Africa* | |
|---|---|
| *Western Rift* <br> Nyamlagira <br> Nyiragongo | *Cameroon* <br> Mt Cameroon |
| *Eastern Rift* <br> Teleki <br> Likaiyu <br> Longonot <br> Meru <br> Oldoinyo Lengai | *Cape Verde Islands* <br> Fogo <br><br> *Réunion* <br> Piton de la Fournaise |
| *Afar Triangle* <br> Erta Ale <br> Gada Ale <br> Dala Filla <br> Ale Bagu <br> Borale Ale <br> Alayta <br> Ayelu <br> Dubbi | *Comoro Islands* <br> Kartala (on Grand <br> Comoro: Ngazija) <br><br> *Canary Islands* – 3 islands <br> have active volcanoes. <br> These are: La Palma, <br> Lanzarote, and Tenerife. |

Plate 3.3 Lake Nyungu lying in an explosion crater near the road between Kichwamba and Rubirizi, SW Uganda

## EXPLOSION CRATER
### (Ring crater)

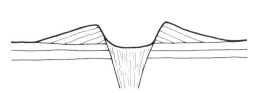

Shallow, flat floored depression surrounded by a low rim of pyroclasts and local rock, usually less than 50 m high. Often found in groups. A vent is blown through the local rock by a series of gas explosions.

## ASH CONES AND CINDER CONES
### (Scoria cone)

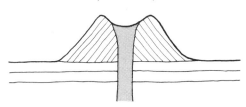

Small symmetrical cones of pyroclasts, usually less than 150 m, with fairly steep sides, 25°–35°. Craters are large and cover all summit area. Cinder cones have steeper slopes than ash cones. Both types often erupt in groups and on the flanks of larger volcanoes The magma has a high gas content and eruption is explosive.

## BASALT DOME
### (Shield volcano)

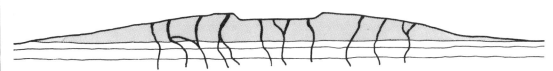

Large flat topped convex dome of basic lava with gently sloping sides, usually low in height relative to a large basal diameter. This shape is caused by very fluid lava able to flow long distances. Large, shallow, steep-sided, sunken crater. Lava tends to flow through numerous fissures rather than a single vent. Small subsidiary cones may form along fissures. Gases are easily released from basic lava and eruptions are therefore not explosive.

## CUMULO DOME          PLUG DOME

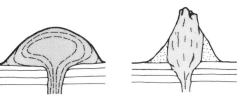

Steep sided convex dome of acid or intermediate lava. This viscous lava does not flow far and piles up around the vent. It hardens quickly and later extrusions, unable to reach the surface may force initial layers outwards. Usually has no visible crater.

Sometimes lava is so viscous it is forced out as a rigid cylindrical plug. The base of the plug is surrounded by exploded debris. Eruptions are very explosive and plugs are extruded amid clouds of hot incandescent ash and cinders.

## COMPOSITE VOLCANO
### (Strato-volcano)

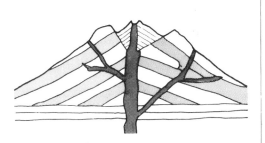

Large cone built of alternating layers of pyroclasts and lava with concave sides at about 20°–30°, often growing to great size, and may have more than one crater. Eruptions are explosive or effusive depending on the gas content of the magma. If the top of the cone is destroyed in an eruption a secondary cone may form. Parasitic cones are common on the sides.

Fig. 3.3 Types of volcano

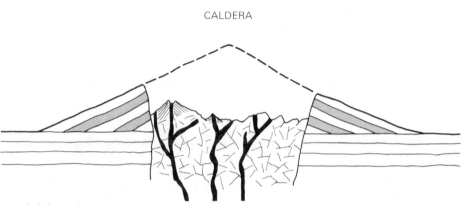

CALDERA

A large rounded depression resulting from the destruction of the upper part of a volcano in a violent eruption. Sometimes the summit cone is blown off and explodes into incandescent rocks and ashes. Calderas are also due to subsidence. Major eruptions, by reducing the magma supply, leave a huge chasm beneath a volcano. The weight of the overlying cone becomes too great, faults develop, and it collapses into the chasm. Many calderas probably result from both explosion and subsidence. Later, new cones may begin to form on the floor of the caldera.

DISSECTED VOLCANO

A volcano with its slopes and crater deeply dissected by radial valleys. Erosion develops first on the upper slopes, and in the early stages of dissection, triangular facets of the original volcano, called planezes, may remain on the lower slopes. Eventually the planezes are also destroyed.

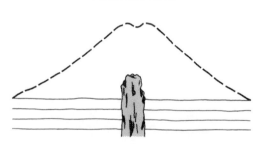

VOLCANIC NECK

A central core of solidified lava in the vent of a volcano exposed by the erosion of the surrounding cone. Besides the volcanic neck, differential erosion may also reveal resistant dykes which have solidified in fissures of the original cone.

Fig. 3.4 Residual volcanic landforms

The world's major regions of volcanic activity are along and near the boundaries of the tectonic plates, especially those land masses near the ocean trenches surrounding the Pacific Ocean, and in the Mediterranean Basin which separates the African and Eurasian plates. These are also the zones of severe earthquakes (Fig. 2.23).

In Africa the distribution of volcanoes is closely related to the pattern of major faults (Fig. 3.5). By far the most active region coincides with the rifts of eastern Africa, but other areas are also important such as the offshore islands of Reunion, Comoro, and the Canaries.

*Some examples of recent volcanic eruptions in Africa.*

| | |
|---|---|
| San Juan (on La Palma, Canaries) | 1949 |
| Fogo | 1951 |
| Kartala | 1952 |
| Mt Cameroon | 1959 |
| Piton de la Fournaise | 1960 |
| Oldoinyo Lengai | 1966 |
| Nyamlagira | 1971 |
| Erta Ale | 1973 |
| Nyiragongo | 1977 |

San Juan's 1949 eruption produced a flow of lava that reached the coast 8 km SW of Los Llanos on the

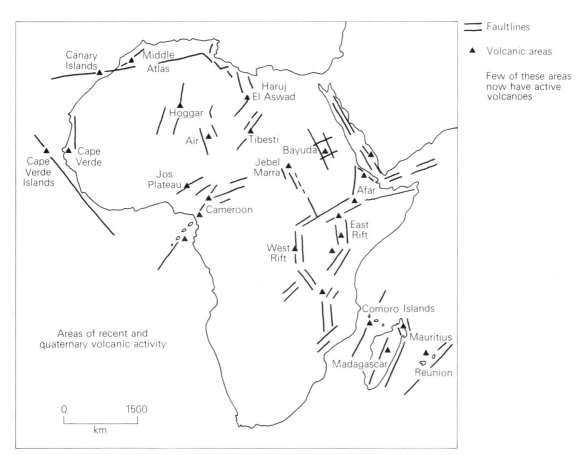

Faultlines

▲ Volcanic areas

Few of these areas
now have active
volcanoes

Fig. 3.5 Areas of recent and quaternary volcanic activity

west coast of La Palma. It protruded into the sea for
1 km forming a brand new section of coastline,
Cabo de San Juan.

*Explosion craters in SW Uganda*
South of Fort Portal, between Lakes Mobutu and
Amin (Edward), there are more than 200 explosion
craters many of which contain shallow lakes (Plate
3.3 and Fig. 3.6). Most are circular, but some
intersect with others to form a more complex plan
(Uganda 1:50,000 sheet 75/2). The average diameter is less than a kilometre, but there are larger ones
such as Lake Katwe, which at its widest is 3 km. The
surrounding rim of Katwe is also higher than most.
It is about 90 m while the average is 45 m. The
craters are Pleistocene in age and were formed
during two periods of activity. They were blown
through the crystalline basement rocks by violent

gas explosions, and today those of the earlier
periods show clear signs of erosion.

Other examples are: near Lake Basotu, 60 km NE
of Singida, Tanzania (e.g. Ghama Crater, Ndobot
Crater), in the Hoggar Mountains, Algeria (e.g.
Ehéri Crater, Ouksem Crater) and near Debra Zeit,
45 km SE of Addis Ababa (e.g. Bishoftu Crater,
Hora Crater).

*Ash cones and cinder cones in Kenya and Nigeria*
South of Lake Turkana, Kenya (1:100,000 Kenya
sheet 53), is a volcanic ridge, the remains of an
ancient volcano, called the Barrier. On both sides of
this ridge are recently active ash and cinder cones
(Fig. 3.7). Likaiyu (Plate 3.4) is a black cinder cone
about 100 m high on the south slopes of the Barrier,
facing the Suguta Valley. Since the formation of the
cone lava has also been erupted causing a breach in

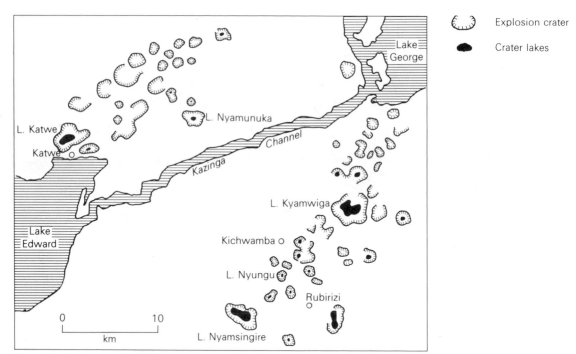

Crater lakes

Fig. 3.6 Explosion craters in SW Uganda

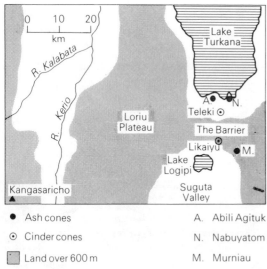

• Ash cones     A. Abili Agituk

⊙ Cinder cones     N. Nabuyatom

▨ Land over 600 m     M. Murniau

Fig. 3.7 Cinder cones and ash cones south of Lake Rudolf, Kenya

the lip of the oval-shaped crater. On the north of the Barrier is Teleki, a large cinder cone that has been fractured by fissure eruptions of lava. Much of the Barrier is covered by a layer of lava and cinders through which project several ash cones. The largest of the cones, built of yellow coloured ash, which contrasts with the surrounding black lavas, are Nabuyatom, Murniau and Abili Agituk. The latter two are much older than Nabuyatom and have been extensively eroded by gullies.

On the Jos Plateau of Nigeria (1 : 100,000 Nigeria sheet 190), there are a number of small cinder cones south of Panyam. They are all extinct and their craters have been breached by later eruptions of basaltic lava. The steep-sided cones are about 100 m high and aligned along a north–south fracture zone (Fig. 3.8).

Ash and cinder cones are often the result of one short-lived period of eruption, for example the 80 m high cinder cone of Kitsimbanyi, north of Nyamlagira in Zaire (Fig. 3.9). Kitsimbanyi means 'one who builds quickly', and was formed in 1958. Other examples are Busoka, Bitale, and others between Muhavura and Lake Mutanda, SW Uganda (Fig. 3.9); Sarabwe and Fileko, 18 km NE of Tukuyu, Tanzania; Mathaioni, Sambu and hundreds of

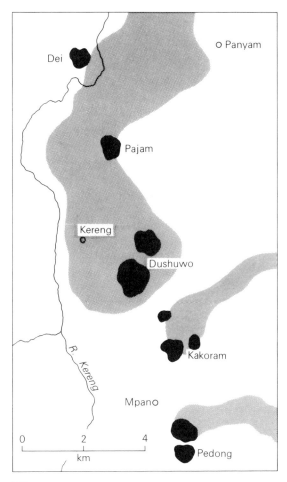

Lava flows

Cinder cones

Fig. 3.8 Cinder cones near Panyam, Nigeria, 65 km SE of Jos

Plate 3.4 The cinder cone of Likaiyu viewed from the north with the Suguta valley in the background. In front of Likaiyu is an exploded cone and a large sheet of black lava

above its base to a maximum height of 3056 m. It is therefore over 300 m lower than its neighbours, and has a typically large basal diameter of 21 km. The sides of the volcano slope gently at about 7° up to a vast steep-walled sunken crater over 2 km wide and 75 m deep (Plate 3.5). The crater was formed by the subsidence of the volcano summit. Its floor is not flat but appears as a series of descending levels, ending in deep pits or lava sinks which contain molten lava.

Nyamlagira has been built up by the successive outpourings of enormous lava flows. The most notable recorded eruption was that beginning in 1938 which resulted in the outflow of about $2\frac{1}{2}$ cubic kilometres of basaltic lava (Plate 1.4). As is typical of basalt domes, most eruptions have been from fissures and subsidiary vents on the sides. It was the removal of lava in this way that caused the volcano summit to subside and produce the vast crater. The flanks of Nyamlagira are covered by a multitude of lava flows and small parasitic cones.

Other examples are: Piton de la Fournaise on Réunion island (Fig. 3.15 and Plate 3.2); Kartala, Grand Comoro island; Erta Ale, Mat 'Ala and Alayta, near Lake Afrera (L. Giulietti) in the Afar Triangle, Ethiopia (Fig. 2.4).

### Cumulo domes in the Itasy Massif, Madagascar

The volcanic massif of Itasy is Quaternary in age and lies 75 km west of Tananarive (Antananarivo).

others in the Kibwezi area of Kenya (Fig. 2.5), south of the Nairobi–Mombasa railway; Volcan de Tao on Lanzarote, and Tahuya on La Palma, Canary Islands.

### Basalt dome of Nyamlagira, Zaire

Nyamlagira is one of a group of 8 major volcanoes on the Uganda–Zaire frontier, known as the Virunga Range or Mufumbiro Range (Fig. 3.9). The range, which also includes numerous small cones and craters, lies in the Western Rift between Lakes Amin and Kivu. All the main volcanoes except Nyamlagira are composite. Nyamlagira rises 1500 m

41

Plate 3.5 Inside the vast sunken crater of the shield volcano of Nyamlagira, Zaire. This photo was taken after the 1938 eruption

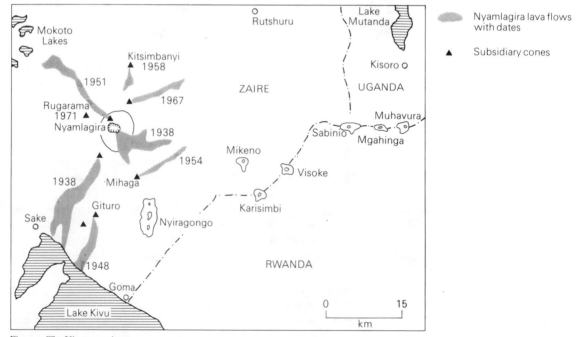

Fig. 3.9 The Virunga volcanoes

Kitia   Ambasy               Antanimarina                        Antsahondra
                                                    Antsakarivo

Plate 3.6 Cumulo domes and cinder cones in the Itasy Massif, Madagascar. The view is looking west from point X in Fig. 3.10. Lake Itasy is in the foreground

It covers over 400 square kilometres and contains large number of small cumulo domes, cinder cones and explosion craters (Madagascar 1:100,000 sheet M47). A significant group of trachyte cumulo domes occurs west of Lake Itasy (Fig. 3.10 and Plate 3.6). The domes are typically without craters and rise to about 300 m above the underlying Pre-cambrian gneiss and quartzite rocks. Some, such as Antsakarivo, have much flatter summits than others, but the majority, especially Ambohibe, show a steep-sided dome shape that has been accentuated by internal swelling from additional lava.

There are more than 60 well-preserved cinder cones in the Itasy Massif. Most are small, about 60 m high, but some such as Kassigie are over 200 m. These cones seem to be the earliest form of volcanic activity in the area, older that is than the cumulo domes. Their slopes are not as steep as the domes, being approximately 25–30°. Many of the cones, which have typically large craters at their summits, have breached walls where lava issued during later eruptions. The lava flows helped form barrier lakes, of which the largest is Lake Itasy (page 235). The explosion craters of Itasy are less numerous than in Uganda. Most of them now hold lakes, the deepest of which is Lake Kitia, being about 100 m. The craters are more recent than the cumulo domes, as is witnessed by the fact that the sides of Kitia and Ambohibe are covered by trachyte and gneiss debris exploded from the Kitia crater.

Other examples of small cumulo domes occur 30 km east of Mbeya, Tanzania, especially Ntumbi, which has no visible crater and is about 250 m high. Cumulo domes that form inside the crater or caldera of a larger volcano are called tholoids. A tholoid, 300 m high, has formed in the caldera of Mt Rungwe, 40 km south of Mbeya, Tanzania.

*Plug Domes in the Hoggar Mountains, Algeria*
The highest part of the Hoggar (Fig. 8.15) is the volcanic district of the Atakor (Algeria IGN 1:200,000 sheet NF-31-XXIV). There are many types of volcano in the Atakor but the most

43

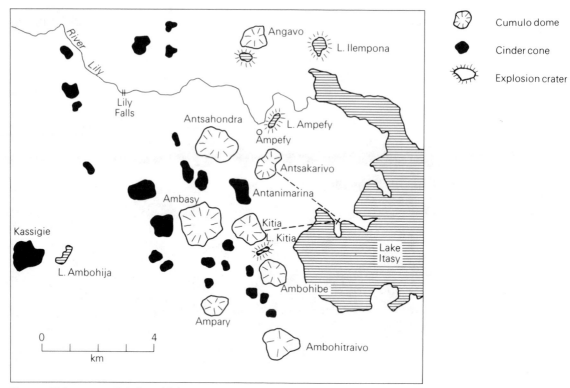

Legend:
- ❀ Cumulo dome
- ● Cinder cone
- ✺ Explosion crater

Fig. 3.10 Volcanoes near Lake Itasy, Madagascar

individual are the plug domes, of which there are more than three hundred. They are composed of phonolite, an intermediate lava (page 34), and rise steeply in a cylindrical shape to 300 or 400 m. Among the most magnificent are Ilaman, Oul and Iharen (Plate 3.7). None of the volcanoes are now active, but they are thought to be the result of eruptions similar to that of Mont Pelée in the West Indies. Initially several explosions took place and blew holes through the Precambrian granites and more recent basalts. This gave rise to the formation of low cones. Through the craters of these cones thick plugs of lava were forced up like giant pistons completely blocking the vents. The spine of Mont Pelée was weak and soon destroyed by erosion, but the Atakor plugs are clearly of a more resistant nature (Plate 8.1). These plug domes are similar in appearance to volcanic necks, and the two types should not be confused.

Other examples are: in the Mandara Mountains,

north Cameroon (e.g. Roumsiki) and on the island of São Tomé, West Africa (e.g. Cão Grande).

*Simple Composite Cone of Muhavura, Uganda*
Muhavura (4129 m) is the easternmost volcano of the Virunga Range (Fig. 3.9) and lies in SW Uganda, on the Rwanda frontier (Uganda 1:50,000 sheet 93/3). It is an almost perfectly symmetrical cone rising nearly 3000 m above its base (Plate 11.7). At the summit is an unbroken crater in which lies a small lake less than 50 m in diameter. On the north-east slopes are some extinct parasitic cones. Muhavura is composed of successive layers of lava and very ashy pyroclasts. The volcano is no longer active and its northern slopes have in places been deeply eroded.

Other examples are: Oldoinyo Lengai, Tanzania (Fig. 2.5), an active volcano almost entirely built of soda-rich carbonate lavas and pyroclasts; Mgahinga, Uganda (Fig. 3.9).

Plate 3.7 Iharen, one of the many extinct plug volcanoes in the Hoggar Mts, Algeria. Iharen, with its strong columnar jointing, lies at the edge of the Hoggar, some 10 km NE of Tamanrasset

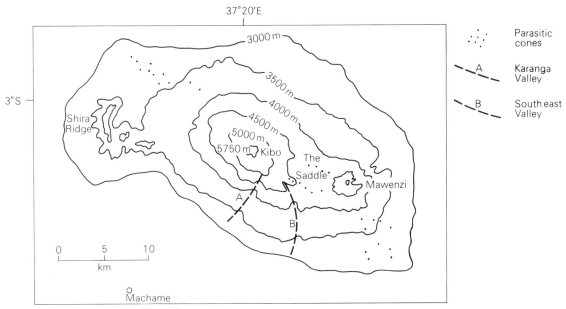

Fig. 3.11 Kilimanjaro

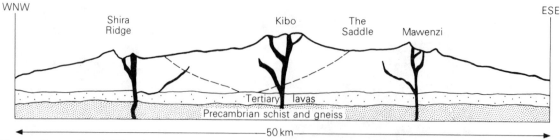

Fig. 3.12 Section through Kilimanjaro

### Complex Composite Cones: Kilimanjaro and Mount Cameroon

Kilimanjaro (5895 m) is a vast volcanic mountain rising 4000 m above the surrounding plains (1:100,000 Kilimanjaro Special Sheet). It is made up of three separate volcanoes which, in order of their formation, are Shira in the west, Mawenzi in the east and Kibo in the centre (Fig. 3.11). Shira and Mawenzi were initially two volcanoes whose peaks were 25 km apart (Fig. 3.12). Today little remains of Shira, except the western rim of a caldera, known as Shira Ridge, while Mawenzi has also been severely dissected (Plate 3.8). Kibo erupted between these two, building itself on their flanks. It is topped by a glacier-covered subsidence caldera, in the centre of which is an inner crater called the Ash Pit (Plate 3.1). More than 250 small parasitic cinder and lava cones cover the flanks of Kilimanjaro, mostly under 100 m high. These cones are especially dominant south-east of Mawenzi (e.g. Kifinika) and north of Shira (e.g. Lagumushira). Volcanic activity on Kilimanjaro is now mainly limited to fumaroles around Kibo's crater. The position of the volcano in a trough on the east side of the Eastern Rift and its proximity to other major volcanoes is shown in Fig. 2.5.

Mount Cameroon (4070 m) is a similar huge volcanic mountain, being 50 km long north-east to south-west and about 30 km wide (Cameroon IGN 1:200,000 sheet NB-32-IV). It now stands alone (Fig. 3.13) as the only active volcano amongst numerous extinct cones and calderas in a zone that extends from north Cameroon south to Macias Nguema Biyoga (Fernando Po) and its neighbouring offshore islands. Fako, the highest point on Mount Cameroon, is on the edge of an extinct crater and volcanic activity is now restricted to occasional

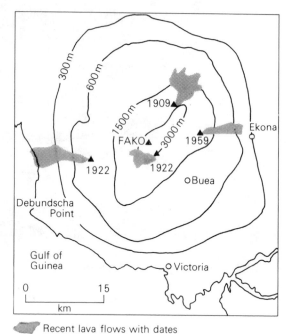

Recent lava flows with dates

▲ Subsidiary vents

Fig. 3.13 Mount Cameroon

eruptions from parasitic cones. In the present century there have been three main eruptions: 1909, 1922, 1959.

Other examples are: Meru, Tanzania (Fig. 2.5); Nyiragongo, Zaire (Fig. 3.9); Emi Koussi, Tibesti Mts. and Pico de Teide, Canary Islands.

### Calderas in Cameroon and Chad

Eboga caldera (Fig. 3.14) is in the Manengouba Massif near Nkongsomba, and lies 120 km NW of Mount Cameroon. The bounding walls are steep and rise 300 m above the caldera floor, which is approximately at 1900 m above sea level. Within

Plate 3.8 Kilimanjaro. A view from the east with the eroded neck of Mawenzi in the immediate foreground and snow-capped Kibo in the background

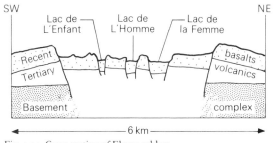

Fig. 3.14 Cross section of Eboga caldera

Eboga are 3 craters, all occupied by lakes. Lac de la Femme is over 160 m deep. There are also many small cones in the caldera, notably Mboroko. On its east side Eboga intersects an older caldera called Elengoum. Fracturing of rocks within the caldera together with only a limited quantity of pyroclasts suggests Eboga was mainly caused by subsidence.

Trou au Natron lies SE of Toussidé volcano in the Tibesti Mountains. It is a steep-sided caldera nearly 8 km in diameter and 1000 m deep. Piecemeal collapse following a number of large-scale explosions would seem to have been the main cause of the Trou. The surrounding slopes are covered by a variety of rock fragments ejected during these explosions, including basement rocks as well as lavas. On the caldera floor, partly concealed by a deposit of carbonate of soda, are 4 small cones (Plate 3.9).

Longonot, in the Eastern Rift (Fig. 2.5), is a good example of a caldera caused by a violent explosion. The flanks of Longonot are dissected by a network of radiating valleys (Plate 3.10, Kenya 1 : 50,000 sheet 133/4).

Other examples are: Menengai and Suswa, in the

Plate 3.9 Trou au Natron, one of several calderas in the Tibesti Mts, Chad

Eastern Rift, Kenya (Fig. 2.5); Ngorongoro, Tanzania (Fig. 2.5); Shala, Ethiopian Rift, occupied by Lake Shala (page 234); Voon, Toon and Yega in the Tibesti Mountains, Chad.

*Dissected volcanoes and volcanic necks*

Réunion (IGN Réunion 1:100,000) is built of two volcanoes, of which the largest is the extinct Piton des Neiges (Fig. 3.15). This volcano is being slowly dissected by three big amphitheatre-shaped valleys: Mafatte, Salazie and Cilaos. These valleys, with their vertical walls and narrow exits, have cut into the volcano flanks to leave only triangular planezes that taper up into sharp-edged peaks. Volcanic rocks eroded from the valleys now form three large debris cones (page 96). The former summit of Piton des Neiges, long since destroyed, was probably 300 m higher than the present peak.

The neighbouring volcanic island of Mauritius (1:25,000 DOS sheet 4) has been even more severely dissected. Jagged uplands, such as Moka-Long Range behind Port Louis (Plate 3.11), are the last remnants of a huge volcano that first erupted in the Tertiary era.

Mount Kenya (Mount Kenya 1:25,000 Special Sheet) is in a very advanced stage of dissection and

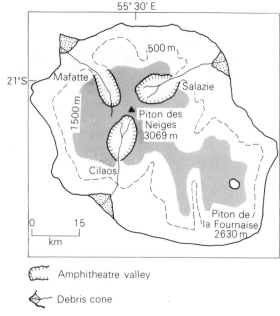

Fig. 3.15 Réunion, the volcanic island east of Madagascar

may once have been nearly 1000 m higher. The main peaks, Batian and Nelion (Plate 9.2), are part of a syenite neck, while the uplands that surround this central core are the remains of the original cone.

48

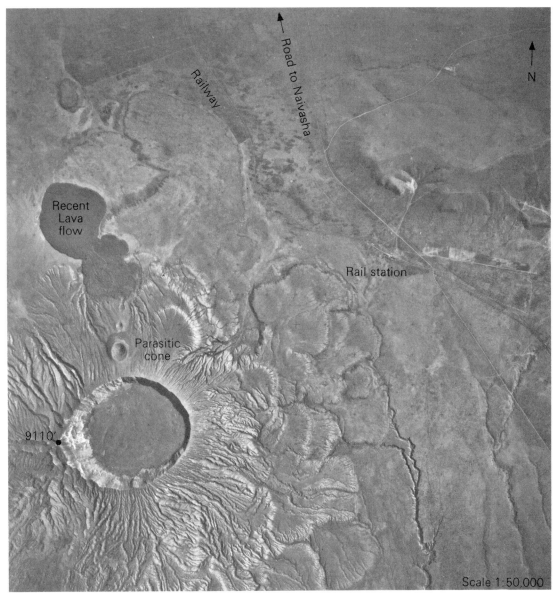

On the image:
- Road to Naivasha
- Railway
- N
- Recent Lava flow
- Parasitic cone
- Rail station
- 9110′
- Scale 1:50,000

Plate 3.10 The circular caldera of Longonot, SE of Lake Naivasha, Kenya, with its flat floor and vertical walls. On the north slope can be seen two small parasitic cones and the dark mass of a lava flow

Wase Rock in Nigeria (Plate 3.12) is a trachyte neck rising 295 m from the floor of the Benue Rift, 160 km SE of Jos. It is all that remains of a once giant volcano. The rock, with its vertical sides, has its summit split in two by a narrow chasm and the base is surrounded by talus slopes (page 62).

Other volcanic necks are: Tororo Rock, SE Uganda; Mawenzi, Kilimanjaro (Plate 3.8); Alekilek (part of Napak volcanic complex) near Lothaa, Uganda; Ehi Goudroussou, north of Bardai, Tibesti Mountains, Chad.

## B. Lava plateaus

A lava plateau is an upland with a generally level summit, and made of successive layers of lava. It is formed by the eruption of very fluid basic lava from

49

Plate 3.11 Pieter Both, the highest peak of the Moka-Long Range, part of an eroded volcano rising behind Port Louis, Mauritius

a large number of linear or fissure vents in the crust. As successive eruptions take place, with little explosive action, very mobile basaltic lava spreads out over preceding flows. Eventually, the depth of lava may be hundreds of metres thick, completely covering any original landscape of hills and valleys (Fig. 3.16). Some lava plateaus are hundreds of square kilometres in area. Vertical jointing in the basalt causes the plateau edges to be very abrupt, and where the plateau has been dissected by rivers the valleys tend to be steep-sided gorges.

*Lava plateaus in Libya and South Africa*
The Haruj el Aswad is a plateau of Quaternary basalts covering about 25,000 sq km in central Libya

east of Sebha. It slopes up gradually to over 750 m above the Libyan Plain and is highest in the north-west (1180 m asl), where steep cliffs overlook the oasis of El Fugha. Most of the plateau surface is rough and covered by low craters and pyroclasts.

The upper part of the Drakensberg Mountains, South Africa, is a lava plateau formed millions of years ago at the start of the Mesozoic era. The lavas must once have covered a large area of southern Africa, but they are now limited to the SE where they form the rim of the Great Escarpment (Fig. 3.17). In places the lavas are over 1200 m thick, covering most of Lesotho and adjacent parts of South Africa. Erosion has destroyed much of the original landscape leaving a dissected plateau at

Plate 3.12 Wase Rock, an ancient volcanic neck on the floor of the Benue valley, Nigeria

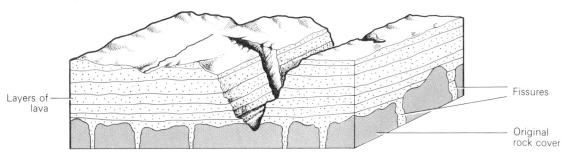

Fig. 3.16 Lava plateau dissected by a deep gorge. Note the fissures along which the lava welled up

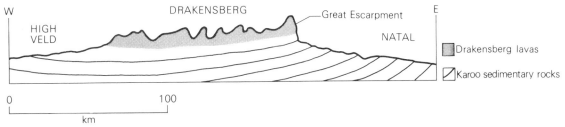

Fig. 3.17 Diagrammatic section of the Drakensberg lava plateau

about 2500–3000 m, above which rise the summits of Giants Castle and Mont aux Sources. The upper tributaries of the Orange River have cut steep-sided gorges, several hundred metres deep, into the plateau, giving rise to mountainous relief in which the vertically jointed basalts form cliffs that resemble castle battlements (Plate 3.13).

In several parts of Africa there have been extensive extrusions of lava, but not always in sufficient quantity to form a plateau, for example

Plate 3.13 The Devil's Tooth, near Mont Aux Sources in the Drakensberg Mts South Africa. The dissected edge of an ancient lava plateau, the top of which is over 3,000 m above sea level

Laikipian Basalts which cover much of the eastern slope of the Aberdare Range in Kenya and the Alayta lava field SE of Lake Afrera, Afar Triangle, Ethiopia (Fig. 2.4).

Other examples of lava plateaus are: central part of the Ethiopian Highlands and Biu Plateau, northern Nigeria.

## Intrusive landforms

The eruption of magma at the surface to form igneous rocks has an immediate influence on relief in the form of extrusive landforms. Some intrusions may form near the surface and cause doming of the overlying rocks, but generally intrusive rocks affect relief only indirectly after they have been exposed by denudation. Whether intrusions form important features in the landscape depends on their relative hardness, for they may be more or less resistant than the surrounding rock, and the effect of erosion will

be to destroy the softer rocks (Chapter 6).

The size and shape of an intrusion depends partly on the nature of the intruded rock (Fig. 1.1), and partly on the structure of the rocks into which the magma is intruded. The depth at which an intrusion was formed can often be determined by the rock type.

The main types of intrusion are shown in Fig. 3.18 with the exception of the largest, the batholith, which is described below.

### Batholith

A very large intrusion of generally granite rock, formed at great depth and apparently bottomless. On exposure at the surface batholiths may form uplands covering hundreds of square kilometres. Some are magmatic in origin, that is they result from the large-scale intrusion of molten rock. This type are of uniform rock and have clear cut margins. Others are due to metasomatism, the replacement of the original rock by mineral and chemical changes. This type has no sharp boundary but grades

DYKE   Vertical or steeply inclined rock sheet intruded into fissures. Thickness varies from centimetres to hundreds of metres. Discordant across rock
strata. May occur in large swarms and have a parallel
or radial pattern. Differential erosion produces a
wall-like ridge or a trench according to whether the
dyke is more or less resistant than adjacent rock.

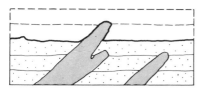

SILL   Tabular rock sheet intruded horizontally into
fissures. Thickness varies from centimetres to hundreds of metres.   Concordant with rock strata. May
occur singly or in large groups. After denudation
resistant sills may form escarpment or flat topped
hills. In river valleys they may cause waterfalls.

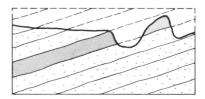

LACCOLITH   Dome-shaped intrusion with flat
floor. Formed of viscous magma which is unable to
spread far and accumulates in a large mass, arching
up the overlying rocks. If the laccolith is more
resistant than the adjacent rocks after denudation it
may form an upland.

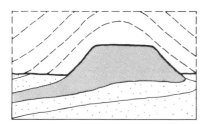

LOPOLITH   Very large saucer-shaped intrusion.
Shape may be due to increased weight causing
sinking. After denudation the upturned edges sometimes form out-facing scarps.

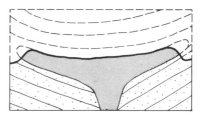

RING COMPLEX   Annular intrusion of one or more ring dykes formed by the process of cauldron subsidence. A
circular block of the crust subsides into an underlying magma vault causing molten rock to rise along vertical fractures
around the block forming ring dykes. Some were formed near the surface, and volcanic activity with caldera formation
preceded fracturing. After denudation they often form a resistant hill mass.

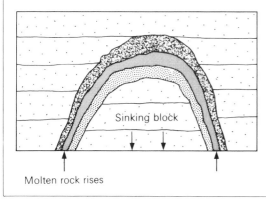

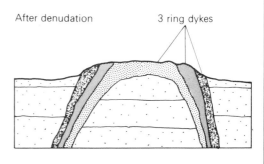

Fig. 3.18 Intrusive landforms

Plate 3.14 A dyke ridge on the SW side of the Kaap valley, South Africa. The Barberton Hills are visible in the background

progressively into metamorphic rocks, hence they have been called metamorphic granites and appear to originate deeper than the magmatic type. The metasomatic type are common in areas of ancient Precambrian rocks. Smaller plutonic rock masses are sometimes intruded around the borders of granitic batholiths, for example pegmatites and aplites.

### Dykes and sills in South and West Africa

In Lesotho, gabbro and dolerite dykes are especially common, varying from 1 m to 10 m in width. When the dyke is intruded into the harder Cave Sandstone, as for example at Lancer's Gap near Maseru, the result is a narrow trench cutting across the pale-coloured sandstone. But when the dyke crosses the soft Molteno Beds the effect is a low ridge, sometimes running straight for several kilometres. In the Kaap Valley, west of Barberton (Fig. 6.5), large numbers of dykes, with a roughly north-western alignment, form vertical ridges in the basement granite. Most of the dykes are about 10 m wide, but on the SW of the valley, near Nelshoogte, the Devils Knuckles and Jambila Dykes form two

wall-like ridges that often exceed 100 m in width (Plate 3.14). Swarms of dolerite dykes outcrop west and south of Blantyre along the Thyolo Scarp, Malawi. They form sharp ridges up to 30 m high between valleys cut in the basement gneisses and schists by the Likabula and other Shire River tributaries.

Sills also outcrop over large areas of southern Africa. Many of them were intruded at the same time as the Drakensberg lavas were extruded, but long-term erosion has left only scattered remnants. They are often to be seen as the protective cap rock on flat topped hills, like Spandau Kop near Graaff Reinet and The Three Sisters in Cape Province (Plate 3.15). In Guinea, dolerite and gabbro sills are the cause of several waterfalls where they form hard rock outcrops along river beds in the Fouta Djallon. Kinkon Falls (Plate 3.16) east of Pita is a good example.

Other examples: west of Lake Turkana (Rudolf) where dykes form linear trenches in sedimentary Turkana Grits, 10 km north of Kangasaricho Hill (Fig. 3.7); in the mountains south of Barberton, South Africa, where rivers have cut valleys across

Plate 3.15 The Three Sisters, South Africa. Three buttes capped by the remnants of a dolerite sill. The hills lie near the Kimberley to Cape Town railway, about 80 km NE of Beaufort West

Plate 3.16 Kinkon Falls, west of Pita, Guinea. Here a headwater of the Konkouré river falls over one of the numerous dolerite sills intruded into the Palaeozoic sandstones of the Fouta Djallon

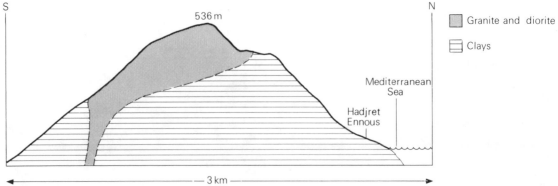

Fig. 3.19 Section through the laccolith of El Koub ou Djaouf, Algeria

quartzite ridges along the lines of dolerite dykes (Plate 6.4); Howick Falls on the Mgeni (Umgeni) River, 20 km NW of Pietermaritzburg, South Africa, which is caused by a 60 m thick dolerite sill.

### Laccoliths in Algeria and Madagascar

El Koub ou Djaouf is a hill over 250 m high lying about 15 km west of Cherchell on the Algerian coast (1 : 50,000 Algeria sheet 39). It is a prominent feature in the landscape due to the fact that the upper part of the hill is a granite and diorite laccolith (Fig. 3.19). The laccolith was intruded into Miocene clays, and although it has since been considerably eroded, forming large screes (page 62) near the base of the hill, it is nevertheless responsible for protecting the underlying weaker clays.

The Fonjay Massif and Ambereny Massif are gabbro laccoliths that outcrop about 30 to 50 km NW of Morafenobe in Madagascar (1 : 100,000 Madagascar sheets F43 & G43). The hard gabbro summit of Fonjay rises 400 m above the sandstones and schists of the adjacent Manambaho lowlands. Ambereny is particularly remarkable for its shape. Although somewhat lower than Fonjay, it covers a larger area and is almost circular in plan with a diameter of about 15 km. The central high point of 530 m is formed from microgranite. Around this lies the main gabbro massif of the laccolith which is in turn surrounded by a wall of more resistant quartzites. From the air, this almost gives Ambereny the appearance of a caldera.

### Lopoliths in Sierra Leone and Rhodesia

The mountainous Sierra Leone Peninsula (1 : 50,000 Sierra Leone sheet 61) is part of a giant lopolith composed mainly of hard gabbro rocks. It differs markedly from the rest of the country geologically and from the neighbouring coastal area topographically (Fig. 3.20). The outcrop of the lopolith is semi-circular, extending from the Banana Islands to

Fig. 3.20 The lopolith of the Sierra Leone Peninsula

Freetown, and coinciding approximately with land over 75 m. The gabbro rocks slope steeply away from the coast to a point that lies several kilometres offshore in the Atlantic. The original area of the intrusion was far larger than that now visible above sea level and may have been as much as 1500 square kilometres.

In Rhodesia, there are a series of 4 large lopoliths intruded into the basement granite and extending across much of the country along a line in a north-easterly direction. The lopoliths have a layered type of structure, the upper part of which is mainly a form of gabbro. The average width of these ancient intrusions is 6 km and they are known collectively as the Great Dyke (Fig. 3.21). Subsequent differential erosion has, for the most part, formed them into a line of hills that reach their highest in the Umvukwe Range north of Salisbury. For much of the distance the hills rise to 200 m above the

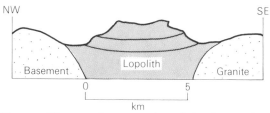

Fig. 3.21 Diagrammatic cross section of the Great Dyke, Rhodesia

adjacent plateau. Near Belingwe, in the south, however, the intrusion forms a zone of lower ground bordered by higher areas of granite.

The Bushveld Basin in the Transvaal of South Africa is also formed of several giant gabbro lopoliths of Precambrian age.

### Ring complexes in Malaŵi and Nigeria

The circular Chambe Plateau is one of a number of ring complexes in south Malaŵi (1 : 50,000 Malaŵi

Plate 3.17 Vertical air view of the giant ring complex of Chambe, Malaŵi

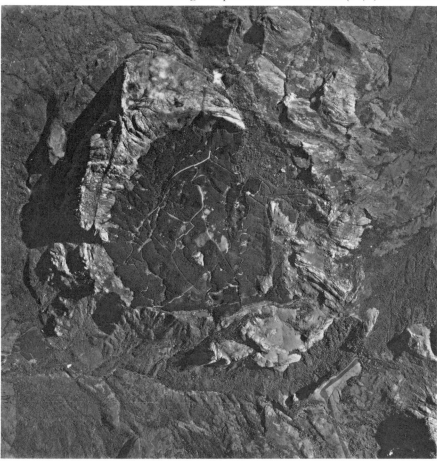

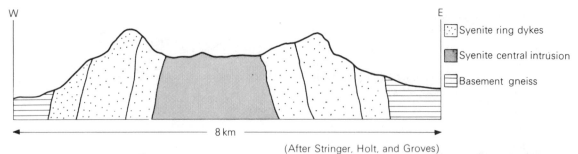

W            E

Syenite ring dykes

Syenite central intrusion

Basement gneiss

8 km

(After Stringer, Holt, and Groves)

Fig. 3.22 Section through Chambe Plateau ring complex

sheet 1535 D3). It lies at the north-west corner of the Mlange Mountains, to which it is connected by a narrow neck into which the headwaters of the Likabula River are slowly cutting back (Plate 3.17). The plateau is built of a central intrusion of syenite surrounded by three large syenite ring dykes. The central intrusion has been eroded down into a crater-like area and the ring dykes form a high encircling steep-sided wall at a height of 1800 m (Fig. 3.22). Since all the rocks are plutonic the complex must have been intruded at considerable depth beneath the surface. The present resistant massif of the Chambe Plateau is the result of long-term differential erosion on the syenites and the surrounding Precambrian gneiss.

The Jos Plateau, rising 500 m above the High Plateau of northern Nigeria, is one of the most prominent relief features in West Africa. The plateau survives mainly because it is composed of a number of hard granite ring complexes making it particularly resistant to erosion. The granites are of Jurassic age intruded into Precambrian gneiss, and because they are more recent than the older basement granites they are known as the Younger Granites. Along the edges of the Jos Plateau the Younger Granites of the ring complexes form steep escarpments, especially in the south-west overlooking the Benue Valley (Plate 7.6). Among the larger of the ring complexes in and around the plateau is that forming the Kudaru Hills (1:50,000 Nigeria sheet 125 SE). The hills, 80 km NW of Jos, correspond almost exactly to the outline of a large oval-shaped ring complex. Covering more than 150 sq km, this complex (Fig. 3.23) has a fairly simple structure – a quartz-porphyry ring dyke surrounds a zone of basement rocks in the middle of which is a central granite intrusion. The granite outcrops as a

central hill mass above the basement complex lowland, around which the ring dyke forms a range of peripheral hills rising 250 m above the plains at their maximum in the north. The peripheral hills are

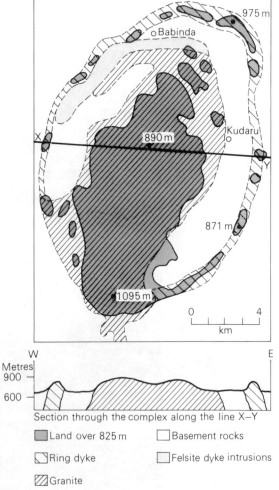

Section through the complex along the line X–Y

Land over 825 m

Basement rocks

Ring dyke

Felsite dyke intrusions

Granite

Fig. 3.23 Kuduru Hills ring complex

Plate 3.18 Granite boulders lying on top of the exposed surface of part of a large batholith, Matopo
Hills, south of Bulawayo, Rhodesia

not continuous but are breached by streams draining radially from the central upland. In the south-
east the granite intrusion abuts directly onto the
ring dyke, with the result that the central upland
and outer ring of hills join and break the inner
lowland in two. In the south-east also the outer
slopes on the ring dyke are especially steep. South of
Babinda, felsite dyke intrusions have formed a
group of minor hills in the inner lowland. (Felsite is
an igneous rock intermediate between rhyolite and
trachyte, Fig. 3.1.)

Other examples are: Sabaloka Hills, Sudan (Plate
5.16 and Fig. 5.42); Tamgak, Agalak and several
others in the Air Mountains, Niger; Zomba
Mountain, Malaŵi.

### Batholiths in Gabon and Rhodesia
Many of the batholiths that outcrop in the worn
down basement rocks of Africa may once have
formed the cores of ancient mountain ranges. Such

batholiths rarely stand out as major uplands today,
for over the millions of years of geological time they
have undergone repeated periods of erosion. Also,
the surrounding metamorphics tend to have similar
mineral and chemical characteristics to the granites
with the result that differential erosion has little
effect.

Some batholiths, however, do produce uplands.
Notable among these is one in southern Gabon
where Precambrian granite outcrops over several
thousand square kilometres and forms the upland of
the Chaillu Massif (Fig. 3.24). This upland, which
rises to 1500 m in Mount Iboundji, extends south-
east into Congo, and is an important watershed
between rivers draining west to the Atlantic and
those draining to the Congo. The massif is covered
by thick forest and has been dissected by numerous
narrow valleys, but in several places bare granite
rock outcrops as inselbergs (Chapter 7) and piles of
massive boulders.

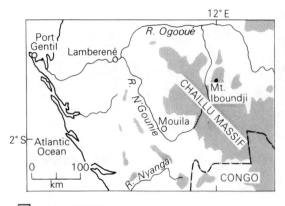

Land over 500 m

Fig. 3.24 Chaillu Massif, Gabon

A large part of Rhodesia is built up of a collective group of several ancient batholiths, partly separated from each other by narrow zones of Precambrian schists. They extend roughly from the Matopo Hills near Bulawayo to the Mtoko area NE of Salisbury. The granite of these batholiths outcrops widely over the country in the form of inselbergs (Chapter 7) and low domes topped by boulders (Plate 3.18).

Other examples are: Tanganyika batholith, outcropping over much of the country between Mwanza and Iringa; Singo batholith, Uganda, a small batholith outcropping between Kikandwa and Kawungera, 70 km NW of Kampala; Sinda batholith, east Zambia, a huge structure outcropping between Petauke and Chipata and extending south into Mozambique; Cape Coast batholith, Ghana, extending approximately from Takoradi NE to Koforidua.

## References

BATTISTINI R., 'Le Massif Volcanique de l'Itasy', *Annales de Géographie*, No. 384, pp. 167–78, 1962.

BATTISTINI R., 'Le Volcan Actif de la Réunion', *Revue de Géographie, Madagascar*, No. 9, pp. 16–44, 1966.

BEDERMAN S. H., 'Mount Cameroon: West Africa's Active Volcano', *Nigerian Geographical Journal*, Vol. 9, No. 2, pp. 115–128, 1966.

GEVERS T. W., 'The 1938–39 Eruption of Nyamlagira Volcano', *Transactions of the Geological Society of South Africa*, Vol. 43, pp. 109–126, 1940.

GÈZE B., 'Les Volcans du Cameroun Occidental', *Bulletin of Volcanology*, No. 13, pp. 63–92, 1953.

GÈZE B, HUDELEY H., VINCENT P, and WACRENIER P., 'Les Volcans du Tibesti', *Bulletin of Volcanology*, Vol. 22, pp. 135–172, 1959.

PUGH J. C., 'The Volcanoes of Nigeria', *Nigerian Geographical Journal*, Vol. 2, No. 1, pp. 26–36, 1958.

STRINGER K. V., HOLT D. N. and GROVES A. W., 'The Chambe Plateau Ring Complex of Nyasaland', *Colonial Geology and Mineral Resources*, London, No. 6, pp. 3–18, 1956.

THORPE M. B., 'The Geomorphology of the Younger Granite Kudaru Hills', *Nigerian Geographical Journal*, Vol. 10, No. 2, pp. 77–90, 1967.

WILCOCKSON W. H., 'Some Aspects of East African Vulcanology', *The Advancement of Science*, Vol. 21, No. 92, pp. 400–412, 1964.

# CHAPTER FOUR

# Weathering and the movement of material on slopes

## Weathering

Weathering is the breaking and decomposition of rocks at or near the earth's surface by physical and chemical processes. The end product is the formation of a layer of weathered rock or debris. The action of weathering in the destruction of rocks, however, is dependent on the removal of the weathered layer by various agents of transport, notably water, wind, ice and mass-wasting. If the weathered rock is not removed it may act as a protective layer to the underlying rock. Thus weathering itself involves no movement of material, but as a process it is dependent upon the movement of material by outside agents. It is an extremely vital process in the shaping of the landscape, first because it is the major means of rock destruction, and second because it is the source on which the erosive agents of wind, waves, running water and moving ice depend for their supply of abrasive rock particles.

Weathering produces many different landforms; for example, karst features (Chapter 6), inselbergs (Chapter 7), rock pedestals (Chapter 8) and rock platforms (Chapter 10).

## Types of weathering

### Chemical weathering

This results in the decomposition of the rock and involves various chemical reactions that take place between the rock minerals, water, and certain atmospheric gases, such as oxygen and carbon dioxide. Also significant are the organic acids that develop from decaying vegetation. Chemical weathering operates to some extent in all parts of the world, but is most active in regions of high temperature and high humidity where chemical reactions take place very rapidly.

Water, charged with different gases, penetrates the rock dissolving and carrying away certain minerals in solution. The chemical nature of the rock is thus changed by transforming the original mineral compounds into new secondary compounds. The main chemical reactions are:

1 Hydrolysis – the reaction between water and mineral elements, that is between the hydrogen ions of water and the ions of the minerals. It is a major process in the decomposition of felspars, important minerals in igneous rocks. The felspars break down to produce potassium hydroxide and aluminosilicic acid. The aluminosilicic acid further decomposes into clay minerals and silicic acid, while the potassium hydroxide reacts with carbon dioxide (present in most water) to produce potassium carbonate. The potassium carbonate is removed in solution, leaving silica and residual clay materials as the end products. The process is depicted thus:

$$2\,KAlSi_3O_8 + 2H_2O + CO_2 \rightarrow$$
felspar        water      carbon
                                    minerals

$$Al_2Si_2O_5(OH)_4 + K_2CO_3 + 4SiO_2$$
clay minerals        potassium    silica
                                carbonate

2 Oxidation – the reaction that occurs when additional oxygen is taken up by a mineral compound. The oxidation of minerals nearly always occurs in association with water, in which atmospheric oxygen has been dissolved. It is most

active in the zone above the water table (Fig. 1.9) and is particularly apparent in sedimentary rocks, such as clays, containing iron compounds. In the saturated zone below the water table the ferrous compounds in the clay give it a grey or blue colour, but in the zone above infiltrating water oxidises the ferrous compounds into red or brown ferric compounds.

3 Carbonation – the process by which carbonic acid (a weak solution of carbon dioxide and water) changes calcium carbonate into calcium bicarbonate which is readily removed in solution by ground water. The process is depicted thus:

$$CaCO_3 + H_2O + CO_2 \rightarrow Ca(HCO_3)_2$$

calcium    water    carbon    calcium
carbonate                dioxide bicarbonate

Carbonation plays its main role in the decomposition of calcareous rocks, especially limestone and dolomite (pages 123–128).

4 Hydration – a process in which certain minerals absorb water and expand causing internal stress and fracturing of the rock. Examples include the conversion of hematite to limonite. Many minerals partially decomposed by other chemical reactions are especially susceptible to hydration (e.g. felspar and mica). In fact, hydration usually occurs in conjunction with other processes, particularly hydrolysis.

These chemical processes, although considered separately, are all closely inter-related and dependent on each other. By their combined action the mineral structure of a rock is weakened, so that eventually the rock decomposes and breaks down into a rotted layer of generally fine-grained debris.

## Physical weathering

This results in the breaking down of the rock into successively smaller fragments and particles. The main processes involve temperature changes. Large daily temperature ranges result in alternate expansion and contraction of certain minerals which cause stress in the rock. Eventually this leads to fracturing of the rock. Further, since rocks are very poor heat conductors, no heat from the outer part of the rock is transferred to the inner part. The outer layers thus expand more than the centre which leads to strains and subsequent fracturing within the rock. How-

ever, it is unlikely that such fracturing will develop without the rock first being weakened by chemical weathering.

In areas subject to alternate freezing and thawing, that is where the temperature fluctuates either side of O°C, water seeps into cracks in the rock and, on freezing, expands and widens the cracks. In well-jointed rocks this can be very potent and in extreme cases the pressure on the rock, due to the increasing volume of the water as it freezes, may exceed 140 kg per sq cm (2000 lb per sq in.).

Expansion and fracturing is caused also by salt crystals penetrating rock crevices, especially in deserts where high evaporation encourages the growth of salt crystals on many rock surfaces.

Physical weathering leads to the disintegration of the rock into blocks (block disintegration), smaller angular fragments, and coarse grains (granular disintegration). The coarse end products of physical weathering contrast with the generally fine debris from chemical weathering. Extensive slopes of angular waste are called talus or scree. Talus slopes are especially common at the base of steep rock faces in desert and mountain areas (Plate 4.1).

## The interdependence of chemical and physical weathering

The essential difference between chemical and physical weathering is that in the case of the latter no actual change is involved in the nature of the rock. The rock is weathered into smaller fragments without any chemical change occurring. Chemical weathering reaches greater depths than physical, which appears to depend more on the removal of the weathered layer.

However, it is vital to remember that in general both types of weathering operate together and are usually complementary, although in a given area one may be more important than the other. Chemical decomposition weakens the rock opening the way for physical disintegration, which in turn, through fracturing, allows deeper penetration by chemical action.

## The weathering front

The weathering front or 'basal surface of weathering' is the sharp divide (Fig. 7.4) or narrow zone separating weathered and unweathered rock. It is generally more apparent in the humid tropics than

Plate 4.1 Devil's Mountain, part of the Acacus Massif north of Ghat in SW Libya. Extensive slopes of talus or scree line the foot of the mountain cliff. In the foreground the physical weathering of the desert has produced sharp-edged angular rock fragments

elsewhere. In many rocks, however, there is a gradual transition zone between weathered and fresh rock, rather than the clear-cut junction of the weathering front.

## The rate and character of weathering

Over much of Africa the weathered layer is often very thick, especially in the tropics where the depth of the weathered zone is roughly 30–45 m (Plate 4.2). This is partly because Africa was not affected by the Pleistocene ice sheets (Chapter 9) which removed most of the weathered rock in northern latitudes. But more directly it is due to the fast rate of chemical weathering in areas of high temperature and humidity. It has been estimated that the rate of weathering in humid tropical Africa is more than twice that operating in temperate countries.

Four main factors affect the rate and character of weathering: the nature of the rock; the climate; plants and animals; relief.

*The nature of the rock*

1 Mineral composition: the mineral composition of rocks is very variable, therefore the character and rate of weathering varies from rock to rock.

Igneous and metamorphic rocks are suscep-tible to chemical weathering because they were formed under temperature and pressure con-ditions very different from those operating at the earth's surface. They are therefore not chemically stable at the surface. On exposure to the atmos-phere chemical reactions set in and change the original minerals into secondary ones. The rate of such changes is faster when acids are present in the water (e.g. carbonic acid from the solution of carbon dioxide in rainwater). The weathering of

63

Plate 4.2 A rail-cutting west of Akroso in the forest zone of southern Ghana exposing over 25 m of deeply weathered granite. This section of the Central Line crosses the Cape Coast batholith from Achiasi to south of Nsawam

one mineral alone is sufficient to weaken the rock, especially in large-grained rocks like granite, since the larger the mineral being weathered the more effective will be the weathering. However, this does not mean that granites weather rapidly, since many other factors influence the speed of weathering.

Some minerals, such as quartz, are stable under atmospheric conditions and are not so susceptible to chemical action. Thus rocks with a high quartz (silica) content, that is acid rocks such as granite, tend to be more resistant than basic rocks such as basalt which have a lower silica content. But the chemical stability also depends on the climate. Under hot tropical climates, in particular, silica is less stable and more easily destroyed.

Sedimentary rocks are more stable than ig-neous and metamorphic rocks since most have been formed by the deposition of previously weathered material which has already undergone chemical change. But a major weakness in all sedimentary rocks is the cement that binds the grains together. Some hard sandstones, for example, are cemented by iron compounds which can be chemically weathered by oxidation (page 61). The iron compounds are oxidised into a brown crust that generally soon crumbles away, although sometimes it is hard and forms a protection against further weathering. Rocks having a calcite cement (the crystalline form of calcium carbonate), for example calcareous sandstone, are attacked by carbonation. One of the strongest cements binding rock particles together is silica. The hard quartz grains cemented by silica make

Plate 4.3 Frost weathering at high altitudes, Nelion Peak, Mount Kenya. Note the joints in the rock face behind the climber's hut and the abundance of angular rock fragments in the foreground

quartzite one of the hardest of rock types and one of the most resistant to weathering.

Most minerals expand when heated, but they expand at different rates according to their composition, size, and colour. Thus light-coloured minerals (e.g. quartz) heat up more slowly than dark (e.g. olivine). The differential expansion and contraction causes stress within the rock which leads eventually to disintegration. Such physical weathering affects igneous and metamorphic rocks which contain a variety of minerals. Internal fracturing is also caused by the chemical process of hydration (page 62).

2 Planes of weakness: planes of weakness are a vital means by which chemical and physical agents can penetrate the rock mass. The most important are joints (page 6), but there is also the bedding in sedimentary rocks and the foliation in metamorphics. Freeze-thaw action is especially dependent on these planes of weakness. The degree of rock jointing is often the major factor influencing the rate of freeze-thaw weathering, and in well-jointed rocks such weathering becomes very effective in the high mountains of East Africa (Plate 4.3).

Many igneous rocks are first attacked by chemical action in the joints. The rectangular blocks of many inselbergs (Plate 7.8) clearly show the influence of joint control. The rounded edges to most of the blocks are due to chemical action of spheroidal weathering, which causes the outer layers of the rock to expand, decay, and crumble. Much of this destruction occurs before the rock is exposed at the surface, a process known as deep weathering.

The combined effect of temperature change and chemical decomposition working along the curved joint systems that result from pressure release (page 6) cause outer parts of the rock to spall or peel off in huge shells (Plate 4.4). This is called exfoliation and it is common on many of the dome-shaped inselbergs that characterise the African landscape, as well as on the slopes of larger structures such as the Chambe Plateau, Malaŵi (page 57).

### The climate

Variations in climate cause differences in the rate and type of weathering. The main climatic controls are temperature and humidity. The climatic regions outlined below are necessarily very broad and hence some generalisation is inevitable.

1 Equatorial regions: chemical decomposition is very active in these latitudes due to high temperatures and rainfall totals. Chemical reactions depend on moisture, and for every increase of $10°C$ the rate of chemical weathering increases 2 or 3 times. This fast rate of decomposition is made evident in the great depth of weathered rock in equatorial Africa, sometimes exceeding 50 m.

The screening effect of the rain forest together with the thick layer of waste debris limit physical weathering in this climate. Temperature change may contribute to the exfoliation of inselbergs

(Chapter 7) and other exposed outcrops, such as when rain suddenly cools a previously heated rock surface, but physical disintegration is not a major factor where humidity is high and both daily and annual temperature ranges are low.

2 Savanna regions: the alternating wet and dry climate of the savannas means that both chemical and physical weathering are important. During the long dry season the sparse vegetation of grass and shrubs gives only limited protection to the ground against changes in temperature and humidity (Plate 4.5). By the end of the dry season the ground surface may come to resemble that of the desert, especially in regions where annual bush burning is common. On bare rock surfaces daily temperature ranges can be as much as 30°C, helping to promote exfoliation and block disintegration. Chemical weathering is especially active during the wet season when humidity is higher, while the high evaporation rates of the dry season help in drawing mineral salts to the surface by capillary action. A major result of the weathering processes in savanna lands is the development of lateritic duricrusts (page 68).

Plate 4.4 Exfoliation on the giant domed inselberg of Ado Rock, near Iseyin, Nigeria

Plate 4.5 Dry savanna landscape near Navrongo, northern Ghana. Gully formation is taking place in the foreground

3 Desert regions: in deserts the extreme daily temperature ranges, low rainfall, high evaporation and the absence of vegetation all greatly increase the effectiveness of physical weathering. In parts of the north Sahara in July the surface temperature on rocks may reach 70°C. The influence of physical weathering is seen most clearly in the occurrence of split pebbles and the angularity of boulders, due to fracturing caused by alternate expansion and contraction. The growth of salt crystals, encouraged by high evaporation, promotes salt weathering on many surfaces (page 62). Moisture seeping into the rock carries dissolved salts with it, weakening the rock and leading to the formation of small cavities called taffonis.

However, it seems likely that chemical weathering, involving moisture drawn from rocks by capillary action as well as from night dews and the occasional rainstorm, is equally if not more important than physical weathering. Chemical decomposition is especially effective in the destruction of crystalline rocks and is most active on outcrops shaded from insolation and able to retain moisture for longer periods. High evaporation draws moisture containing iron and manganese oxides to the surface by capillary action. The oxides are deposited on the rock surface as a thin glazed coating called desert varnish, which has the appearance of a shiny, dark brown enamel. Sometimes this is more than $\frac{1}{2}$ cm thick.

4 Mountain regions: frost weathering is active in mountain areas, that is mainly above 4300 m in the tropics and above 3000 m in the subtropics. The product of this weathering is clearly seen in the large talus or scree slopes that mantle the base of cliffs and high peaks on Ruwenzori and Mount Kenya (Plate 4.3). On Ruwenzori large screes line the foot of the lower slopes of Mount Baker south of the Scott Elliot Pass and also the slopes of Mount Stanley above the Stuhlmann Pass (Fig. 9.15). Frost action also operates in the Drakensberg, Ethiopian Highlands, Cape Mountains and Atlas Mountains. Even in the Sahara there is evidence of freeze-thaw on the upper slopes of the Tibesti volcanoes. Frost weathering, combined with other processes, is important in the formation of periglacial landforms (Chapter 9).

In mountains the rate of physical weathering can be fast, but chemical weathering is limited by the low temperatures and relative absence of liquid water. Many chemical reactions are impossible at such cold temperatures, with the notable exception of carbonation. The effectiveness of carbonation is shown in the active decay of rock beneath snow patches, a process called nivation (page 192). This is because carbon dioxide is far more soluble at low temperatures than at, say 25°C, although the actual rate of weathering is considerably reduced.

5 Sub-tropical regions: in the sub-tropical areas of South Africa and NW Africa north of the Atlas Mountains both physical and chemical weathering operate. However, the latter tends to be more dominant, especially in areas of higher rainfall such as the Natal coast. But in contrast to the tropics the rate of weathering in these subtropical regions is much slower and the depth of the weathered layer is much less.

*Plants and animals*

A thick vegetation cover, such as tropical forest, acts as a protection against physical weathering and also helps to slow down the removal of the weathered layer. In deserts and high mountains the absence of vegetation accelerates the rate of weathering. Plants and animals, however, play a significant part in rock destruction, notably by chemical decomposition through the action of organic acid solutions. The acids develop from water percolating through partly decayed vegetation and animal matter.

The movement of animals and insects in the ground, such as rodents and termites, promotes weathering, by loosening and mixing partially weathered rock, increasing the access of oxygen and water to mineral particles, and carrying organic matter down from the surface. Trees and other plants also act as weathering agents by the action of their roots which grow deep into the ground and open up joints.

*Relief*

A high altitude area with steep slopes suffers different conditions of weathering from a low-lying, gently sloping plain. The rate at which weathering takes place is related to the speed at which the weathered layer is carried away. On steep slopes waste material is soon removed and weathering proceeds faster. Steep slopes and bare rock faces are very susceptible to the frost weathering of Africa's high mountains. In low lying districts and areas of gentle relief chemical weathering is generally more dominant, partly because there is a greater tendency for moisture to accumulate, especially where there is already a deep layer of rotted rock.

## Duricrusts

An important feature associated with weathering and landform development in the tropics is the formation of duricrusts. A duricrust or cuirass is a hard crust formation at or just below the ground surface. There are three main types: ferricrete (iron crust), silcrete (silica crust), calcrete (calcareous crust). Ferricrete or lateritic duricrust is the most significant of these. It occurs in many parts of Africa, but is most apparent in savanna areas where it often produces distinctive landforms.

Laterite first develops as a clay horizon rich in iron oxides within the soil profile. Most laterite sheets are 2–5 m thick, but some layers exceed 10 m implying that the laterite develops over long periods of time. The nature of crusts varies from place to place, which suggests that they are probably formed in more than one way. However, ideal conditions for development seem to occur on gently sloping areas and in depressions where regular flooding leads to the weathered layer becoming impregnated with iron solutions. The iron solutions originate from the leaching of rocks and older laterites on surrounding slopes, and are washed into the weathered layer by laterally-moving ground water and surface drainage.

The formation of ferricrete or lateritic duricrust depends on the removal of the topsoil. Once exposed the laterite hardens into a crust due to wetting and drying which washes out the clay and crystallises the iron. An alternating wet-dry savanna climate therefore has an important effect on duricrust formation, while the removal of the topsoil is hastened by destruction of the vegetation and in this context man often plays a part.

Ferricretes have also been identified in equatorial and desert regions. But it is probable that these, for example those in the Hoggar Mountains (Fig. 8.15), developed under a more humid past climate.

Once formed, all duricrusts show considerable resistance to weathering and erosion producing distinctive landforms in many areas, notably tabular cappings on hilltops. Uplands topped by duricrust layers vary from small buttes to larger plateaus (page 118). Along hillsides and valley slopes the edge of a crust forms sharp benches or low cliffs (Fig. 4.1).

In West Africa ferricretes are most widespread in the zone 10°–16°N, and one area where they produce a very distinctive landscape is in the dissected Ordovician sandstone plateau of the Fouta Djallon in Guinea. For large parts of the level plateau surfaces the crust lies beneath a thin, mainly sandy, overlying cover. But towards the edges of the deep, steep-sided valleys that cut through the Fouta Djallon, sheet-flooding (page 72) has completely removed the overburden to expose a crust that is 2 m and more thick. The exposed duricrusts are most apparent in the west, north of Télimélé, where they produce a hard impermeable surface sometimes totally bare or at most covered by a thin grass and shrub vegetation. This surface is known locally as bowal (pl: bowé), a term meaning 'no trees'. In some areas large shallow depressions occur on the surface, while at the edge where dissection is active removal of the underlying weathered debris produces small caves. In the undulating foothill region north-west of the Fouta Djallon, for example near Madina Bové, there are places where the bowal has broken down into a landscape of blocks and rubble.

In Nigeria laterite duricrust forms the cap layer of the small Share cuesta lying NW of Share town, 60 km NE of Ilorin (Nigeria 1:50,000 sheet 202 NW). The cuesta rises to about 360 m while the crust, which is about 3 m thick, is gently undulating and topped by a moderate soil cover. On the upper part of the valley sides the hard brown crust forms benches where tributaries of the Oshin, Oyi and other rivers are cutting back into the cuesta slopes. At lower levels the slopes are littered with lateritic gravels and blocks resulting from the weathering and erosion of the massive crusts (Fig. 4.1).

Extensive ferricrete surfaces occur also in parts of NW Ghana and in the Sula Mountains of Sierra Leone. Goron Dutsi and Dalla Hills at Kano, Nigeria are particular examples of hills capped with a hard crust layer, and many similar ones occur around Kampala, Uganda.

Calcretes are especially common in arid areas, such as the north Sahara and the Kalahari. Between the Kuruman and Molopo rivers SW of Tsabong (Fig. 8.2) the calcrete crust is thicker than 30 m.

Silcretes are more associated with semi-arid regions. In north Botswana silcretes form minor rapids along streams draining the Okavango delta (Plate 5.11).

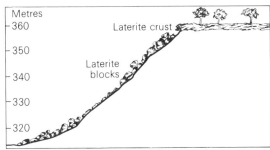

Diagrammatic section of the upper slope of the Share Cuesta, NE of Ilorin, Nigeria.

Fig. 4.1 Diagrammatic section of the upper slope of the Share Cuesta, NE of Ilorin, Nigeria

## The movement of material on slopes

Much of the land surface of Africa is made up of slopes, for the slope, especially the valley side, is the most common landform in the world. The form or profile of any slope is generally a complex grouping of straight (rectilinear) and curved (concave or convex) parts. An actual slope profile will depend mainly on the climate (see page 153), the rock structure (see page 117), and the stage in slope evolution (see page 11).

Slopes, whether they be steep or gentle, originate in a wide variety of ways. Most of these ways, for example faulting, coastal cliff formation, valley erosion by streams and glaciers, moraine deposition, are considered in other chapters. In the present context slopes are important because of the part they play in the processes of mass movement.

Material is moved down slopes by 2 main processes: mass wasting and downwash. These processes are important not so much for the land forms they produce but rather because of the part they play in denudation. By the action of mass wasting and downwash waste material is removed, so facilitating the continuing work of weathering.

### Mass wasting

Mass wasting is the creeping, flowing, sliding or falling of rock and weathered material downhill under the influence of gravity. Gravity can only play its role, however, when the material overcomes its initial resistance to movement. The major factor that helps overcome this resistance is water. A water-saturated mass moves more easily than a dry one because water both increases the weight of the mass and also reduces the cohesion of the materials in the mass.

The main types of mass wasting are shown in Figs. 4.2 and 4.3. It should be noted, however, that there is no sharp break between the various movements, each one overlaps and blends in with the next.

Factors affecting the nature and speed of mass wasting are:

1  The nature of the material and the extent of saturation. Where the weathered layer is very deep, or where rocks are weak, thinly bedded or steeply dipping, wasting will be rapid. Thin beds increase the tendency as there are more bedding planes over which movement can occur. Especially favourable to movement are massive rocks overlying weak rocks, like clay or shales. The more saturated the material, the more likely it is to move.

2  The angle of slope. The steeper the slope, the faster the rate of movement.

3  Climate, including the amount and nature of the rainfall, and the annual and daily temperature ranges. Heavy rain or alternate freezing and thawing encourages movement.

4  The influence of plants. Absence of vegetation to hold the material encourages movement.

5  Man's activities, mining, building, and the herding of animals are among the ways in which

| Type | Material | Slope | Ideal conditions | Occurrence | Speed |
|------|----------|-------|------------------|------------|-------|
| Soil creep | Soil and fine materials | Very gentle | Any process that disturbs soil will cause downslope movement, such as alternate heating and cooling or wetting and drying. A hard pan layer may act as a plane of movement. | Intermittent but sometimes continuous and active over a wide area | Very slow ↕ Slow |
| Solifluction | Saturated soil gravels and weathered rock | Moderate | Limited to mountain and cold climate areas where thawing causes a saturated surface layer to creep as a mass over underlying frozen ground. | | |
| Talus creep | Angular waste rock of all sizes | Moderate | Large talus sheets move en mass especially in mountains where freeze-thaw is frequent. Talus moving down a valley in a long stream is a rock glacier. | Intermittent | |
| Mud flow | Semi-liquid mud with gravels and boulders | Moderate to steep | Large volumes of unconsolidated material, super-saturated after heavy rain, become plastic and flow. Common in arid and semi-arid regions. | and | Fast ↕ |
| Slump | Large masses of rock and debris (Landslides) | Over steepened, e.g. scarps, cliffs and road cuttings | Massive rocks overlying weak rocks saturated by heavy rain. Common on oversteepened slopes. | | |
| Rockslide | | | Surface rocks sliding over a slip-surface formed by bedding or fault planes dipping sharply downslope. | Localised | |
| Rockfall | Individual rocks and boulders | Very steep to vertical | Precipitous slopes in mountains where well-jointed rocks may be loosened by freeze-thaw. | | Very fast |

Fig. 4.2 Major types of mass wasting

man has affected the stability of the surface layers.

6 Earthquakes and volcanic eruptions. These often cause large and widespread movements.

## Downwash

Downwash is the movement of material down slopes by running water as a result of heavy rainfall. The impact of rain splashing on the ground may itself add to the effect of downwash by loosening individual particles. The torrential rainfall of tropical Africa causes downwash to be a very powerful agent of transport. During a heavy storm hundreds of tons of material may be moved. In savanna regions, in particular, the effect of the rainfall is increased by the fact that it follows a lengthy dry season and falls mostly on land that has only a small

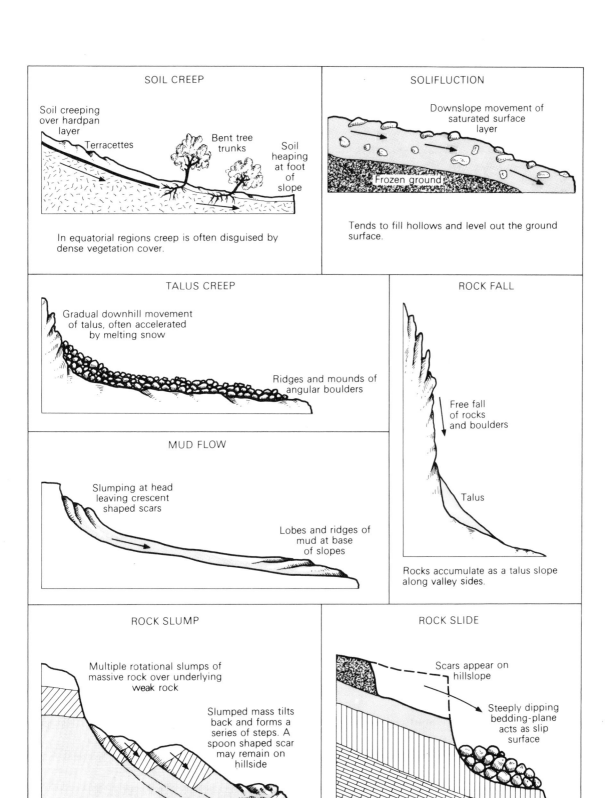

Fig. 4.3  Major types of mass wasting

vegetation cover. There are two main types of downwash.

1 Sheetwash or sheetflooding – this is the removal over a large area of a top layer of soil and other fine material by a thin sheet of water flowing over a fairly smooth surface. In savanna and semi-arid areas sheetwash is especially effective at the start of the rains, when the ground surface is very hard and resistant to percolation. It is particularly effective in washing away the thin cover that overlies lateritic duricrusts (page 68).

2 Gullying – this is the formation of gullies or V-shaped channels in weak materials by the turbulent action of rainwater. Gullying is especially marked on the steep slopes and is caused by irregularities in the ground which concentrate the rain initially into rills or small threads of water which develop miniature channels. Further down slope these rills coalesce and begin to cut gullies, called dongas in southern Africa, which once formed extend themselves headward into the hillslopes. Human activities, such as footpaths, over-grazing, and uncontrolled cultivation, all tend to accelerate the process. As two or more adjacent gullies enlarge themselves and begin to join together they are divided by only a narrow ridge. Eventually the ridge is broken into rows of erosion pillars or earth pillars (Fig. 4.4). Such pillars are formed in weak and soluble sediments, like clay and rock salt. Some may survive for a long time if capped by more resistant material.

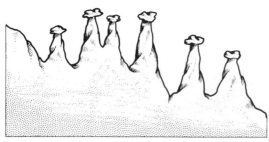

Fig. 4.4 Earth pillars

Similar in origin to gullies but much larger in size are the erosion cirques that dominate some parts of equatorial Africa. Erosion cirques are huge basin-shaped amphitheatres cut out of weak materials by concentrated rain action in conjunction with large-scale slumping. They must not be confused with glacial cirques (page 177). In contrast to gullies, erosion cirques usually have a semicircular shape, are often much bigger and have flatter floors. The largest are generally the result of the coalescence of several smaller ones. This produces a clover leaf plan, with narrow ridges protruding into the cirque.

## Examples of mass wasting and downwash

*Mass wasting in Tanzania, Ghana and Morocco*
Very heavy rains in the Tukuyu area of Tanzania during April and May 1955 were the cause of several landslides (Fig. 4.5). At one time over 425 mm of rain fell in one night. Volcanic ash deposits overlie basaltic lavas in this area, and most of the movements involved the waterlogged ash layers slumping and sliding over the impervious lavas. Landslides were especially prevalent along the steep banks of the rivers, and one on the west bank of the Mbaka was so large, 300,000 cu metres, that it temporarily blocked the river for 8 hours. Most of the movements showed strong rotational slumping in their upper parts, and one at the junction of the Mwalesi and Kandete rivers developed into a mudflow at its base because of the abnormally high water content.

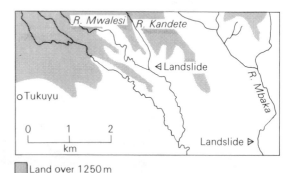

Land over 1250 m

Fig. 4.5 Location of two of the landslides that occurred near Tukuyu, Tanzania, in 1955

Heavy rains in southern Ghana during July and August 1968 were also a cause of numerous landslides, one of the largest of which was that along the Mampong Scarp, north of Kumasi (1:62,500 Ghana

Plate 4.6 Landslide near Jamasi, about 40 km north of Kumasi, Ghana

sheet 131). In this area the thinly bedded Lower Voltaian Sandstones form almost vertical cliffs that rise to nearly 100 m above the road. On two occasions, July 31st and August 23rd, thousands of tons of saturated sandstone slid down and completely blocked the Tamale road near Jamasi (Plate 4.6). Less spectacular slides occurred in other areas, for example along rail-cuttings between Awaso and Dunkwa.

In the Rif Atlas Mountains of Morocco, some 3 km west of Bab Taza, a particularly large mud flow occurred in late February 1963. The flow was in a small tributary valley of the Rhesana River on the northern slopes of Jebel Rhesana. The material was mainly weathered rock and the chief cause of flow was a period of three weeks' continuous rain during which more than 1000 mm was recorded. Movement began on February 26th and continued for

about a week, destroying farms and disrupting drainage. In the upper and middle sections rotational slumping was common. Lower down, depressions were filled with water and formed small temporary lakes. At the very front the flow was characterised by thick tongues of piled up mud which forced the course of the Rhesana River to be displaced to the east by nearly 100 m.

### Ancient lahars in East Africa

The lahar is a special type of mudflow composed of saturated volcanic ash. The most severe lahars are caused by the sudden outflow of a lake in a volcano crater during an eruption. In the past, gigantic lahars have occurred on the flanks of many African volcanoes. The resulting deposits today are often many metres thick and give rise to a hummocky landscape of small mounds, such as south of Mount

Plate 4.7 Tree trunks bent over by creep and slumping on the bank of the Wilge River, near Harrismith, South Africa

Meru near the Arusha–Moshi road (Fig. 2.5), and about 12 km SW of Mount Rungwe, both in Tanzania. On the north and east of Sabinio (Fig. 3.9) in SW Uganda lahar deposits of mud, sand and volcanic boulders up to 8 m thick cover an area of several square kilometres. They have left the scar of a large, deep gully on the upper slopes of the volcano.

### Erosion pillars in Ethiopia

At the north end of the Danakil Depression in the Afar Triangle (Fig. 2.4) vulcanicity has heaved up part of the vast salt plain to form the dome-like Mount Dallol. On the SW of Dallol rain has dissolved and eroded the weak salt into a landscape of huge earth pillars. The higher pillars, some rising over 25 m, are protected by a cap of less soluble gypsum.

### Other examples of mass wasting

Creep is less dramatic and obvious than other movements, but is nevertheless widespread in much of Africa (Plate 4.7).

Solifluction is limited mainly to mountain areas, especially the high peaks of East Africa where it is most marked along moraine ridges (page 182) on the steeper slopes.

Along sea coasts and lake shores rock falls and slumping are often extensive. The Tertiary sandstones and clays of the cliffs between Fresco and Sassandra in Ivory Coast (page 204), especially near Kotrohou, are particularly prone to slumping. On the NW side of Lake Malawi, between Nkhata Bay and Florence Bay, there has been considerable slumping of Precambrian gneisses and schists. (See also Plate 4.8.)

Talus creep is frequent in the high mountain

74

most marked examples is in east Nigeria, where more than 95 km of the Udi Cuesta and adjacent Awka Uplands are cut by gigantic gullies (Fig. 6.7). These gullies have been encouraged by vegetation clearance and violent rainstorms, but they are most extensive in areas where steep slopes coincide with the weak False Bedded Sandstones (Fig. 6.8), and especially the Nanka Sands. The most severe area is near Agulu, Nanka and Oko, east of Onitsha on the edge of the Awka Uplands (Fig. 4.6). Here gullying, sheetwash and slumping have combined to form huge gullies, some of them 150 m deep and over 2 square kilometres in area (Plate 4.9). They have fairly flat floors which are mostly covered with debris slumped down from the steep sides (1 : 50,000 Nigeria sheet 301 SW). Except during the wet season they are dry for much of the year. Such gullies have not formed in the south of the Udi Cuesta, near Okigwi where the more resistant Awgu Sandstones outcrop.

Plate 4.8 Rockfalls in the steep cliffs of Tertiary sediments near Tangier on the north Moroccan coast

areas. Several examples of rock glaciers occur in the head parts of the Taouchguelt and other eastern valleys of Bou Naceur Massif 80 km SE of Taza in the Moroccan Atlas. These features are like giant rivers of rock extending down valley to about 2250 m, and are thought to have formed originally during the late Pleistocene period.

Along the Bahati Scarp, east of Nakuru, Kenya, between Kariandusi and Mbaruk, rock slides of compacted volcanic pyroclasts have, in the past, moved over the finer-grained underlying clayey volcanics.

## Gullies in Eastern Nigeria

In many parts of Africa, from Lesotho to Morocco, gullying has become a serious problem causing wholesale destruction of farmlands. One of the

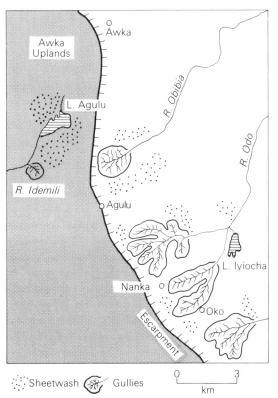

Fig. 4.6 Sheetwash and gullies in the Agulu–Nanka area of East Nigeria

Plate 4.9 A view looking up towards the head of one of the huge gullies near Agulu, Nigeria. (Note the children lower left.)

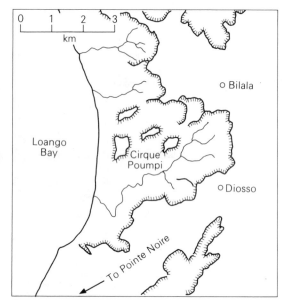

Fig. 4.7 Erosion cirques on the Congo Coast north of Pointe Noire

## Erosion cirques in Congo and Zaire

North of Pointe Noire in Congo (1 : 200,000 IGN sheet SB-32-VI) several very large erosion cirques have been cut out of the weak Pliocene sands that form the coastal plateau, notably Cirque Poumpi at Diosso (Plate 4.10 and Fig. 4.7). The initial development of these features resulted from vegetation clearance, since without cover the soft sands have no protection against the heavy rainfall. Annual rainfall is generally between 1000 mm and 1500 mm, but some years greatly exceed this, for example 1961 when 2048 mm was recorded at Pointe Noire. Over a long period of time small gullies have gradually been enlarged into these huge amphitheatres. The sands at the lower part of the cirque walls are constantly saturated by water seeping from small springs. This promotes slumping and helps maintain the steep walls with their sharp upper edge. During heavy rains the higher slopes become

unstable and collapse, producing huge mudflows. All the fallen material is eventually washed away by rain. Thin gravel beds in the sand are responsible for projecting ledges along some slopes and the long knife-edged crests, topped by earth pillars, that strike away from the steep walls. As the cirques are enlarged forest vegetation begins to invade the floors, colonising those areas no longer affected by downwash and mass wasting. From the air the cirques appear as bright splashes of red, orange and yellow that contrast sharply with the dark green of the vegetation. Other large cirques in Congo include Trou de Gendarme near Mouyondzi and Trou de Dieu near Kinkala (Fig. 5.47).

In Zaire some particularly huge cirques have formed around Masi Manimba and Kitwit. Gouffre de Lukwila (Plate 4.11) near Kandale, 125 km SE of Kitwit, has been cut from the clayey sandstone of the Kwango Series. It is more than 10 sq km in area and over 200 m deep.

In Madagascar, similar erosion amphitheatres, known as 'lavakas' gash many hillslopes in the central highlands, especially south of Lake Aloatra and in the region west of Tananarive, near Miarinarivo and Soavinandriana.

Other examples of serious gullying occur near Maseru, Quthing, and Mafeteng in Lesotho; in

Plate 4.10 The clover leaf shape of Cirque Poumpi from the SW, one of many giant erosion cirques, known locally as Gorges de Diosso, on the Congo coast between Pointe Noire and Tchissanga

parts of Queen Elizabeth National Park, Uganda; and near Ikonongo, 5 km NW of Iringa, Tanzania.

## References

BISHOP W. W., 'Gully Erosion in Queen Elizabeth National Park', *Uganda Journal*, Vol. 26, No. 2, pp. 161–165, 1962.

DE CHÉTELAT E., 'Le Modelé Latéritique de l'Ouest de la Guinée Française', *Revue de Géographie Physique et de Géologie Dynamique*, Vol. 11, pp. 5–120, 1938.

FLOYD B., 'Soil Erosion and Deterioration in Eastern Nigeria', *Nigerian Geographical Journal*, Vol. 8, No. 1, pp. 33–44, 1965.

HALDEMANN E. G., 'Recent Landslide Phenomena in the Rungwe Volcanic Area', *Tanzania Notes and Records*, No. 45, pp. 3–14, 1956.

RIQUIER J., 'Les Lavaka de Madagascar', *Bulletin de Géographie d'Aix-Marseille*, Vol. 69, No. 4, pp. 181–191, 1958.

SAUTTER G., 'Essai sur les formes d'érosion en "cirques" dans la région de Brazzaville, Congo', *Mémoires et Documents, Nouvelle Série*, Vol. 9 (Editions du Centre National de la Recherche Scientifique). 1970.

SHISHIRA E. and WAIGANJO J., 'Itamba Gullies, Iringa', *Journal of the Geographical Association of Tanzania*, No. 4, pp. 21–40, 1969.

THOMAS M., *Tropical Geomorphology*, Macmillan, 1974.

TRICART J., *Landforms of the Humid Tropics, Forests and Savannas*, Longman, 1972.

WIGWE G. A., 'The Laterite Landscape of the Share Area of Ilorin, Nigeria', *Journal of Tropical Geography*, Vol. 38, pp. 61–78, 1974.

Plate 4.11 Part of Gouffre de Lukwila, a huge erosion cirque near Kandale in the Kitwit region of Zaire. Note the earth pillars rising from the floor to the left

# CHAPTER FIVE

# River valleys and drainage patterns

## The work of rivers

Rivers are the major agents of land sculpture throughout most of the world. As they flow they carve valleys, transporting and depositing the material they have eroded sometimes hundreds of kilometres from its source. The material transported by a river is known as the load.

The effectiveness of rivers in their work of erosion, transport, and deposition depends on their energy. The greater the energy the greater the ability to erode and transport.

River energy depends mainly on the volume and speed of the water. The greater the volume the greater is the energy, and, except in very dry areas, volume increases from source to mouth. The speed of the water depends on the gradient of the stream channel. The steeper the gradient the faster the speed and the greater the energy. But speed is also influenced by friction of the water against the sides and bed of the channel, hence speeds are generally greater in large deep channels where such friction is limited. This is apparent in the lower course of rivers where a larger channel counteracts the possibility of any reduction in speed caused by a decreasing gradient. In addition, the volume of water in the lower course is generally increased from tributaries. Such factors explain why the average speed of water in a river tends to increase from source to mouth. Energy is especially great during floods when both the volume and speed of water are increased enormously.

In Africa, outside the equatorial region, there are great seasonal variations in volume. In fact, most African rivers are dry for much of the year and in flood when the rains come. The water discharge of the Niger is 30,000 cu metres per sec in the wet season, but only 1,200 cu metres per sec in the dry season. The energy of a river, therefore, varies from place to place and time to time.

### River transport

The larger the river the greater is the amount of energy available for transporting the load. Although it must also be remembered that a large river will have a correspondingly large load to transport. The Zambesi, for example, discharges 100 million tons of sediment each year.

A river's load includes weathered debris from the valley sides and rock particles eroded from the bottom and sides of the channel. The size of the load varies according to the volume and speed of the water. During floods the great increase in discharge makes it possible for rivers to move much larger loads, that is, their carrying capacity is greatly increased (Plate 5.1).

The load-carrying ability of the river is also indicated by its competence. Competence is defined as the largest particle the river can move, and it varies with the speed and volume of the water. Only during floods are rivers competent to move very large boulders. The size of the load at any point also depends on the geology and climate of the area through which the river is flowing.

Stream load is important since it affects the rate of erosion. Too great a load reduces the erosive power and leads to deposition. As the load is carried downstream, the materials are gradually reduced in

Plate 5.1 Giant boulders in the bed of the Bujuku river, Ruwenzori. The glaciated peak of Mt Stanley is visible in the background

size by striking against each other and against the channel sides. This process is called attrition.

Rivers transport their loads in four ways:

1. traction – the rolling of large particles and boulders along the bed;
2. saltation – the bouncing of small particles along the bed;
3. suspension – the movement of very fine particles held up by the turbulence of the water;
4. solution – the movement of materials dissolved in the water.

### River erosion

The energy available for erosion in most rivers is very small due to the high percentage used in overcoming friction on the sides and bottom of the channel. The amount of friction depends on the shape of the channel. Wide, shallow channels involve a greater loss of energy through friction than narrow deep channels (Fig. 5.1). Friction

against a rocky, uneven stream bed also reduces energy, as does internal friction of water due to turbulence or eddying in the stream.

River erosion involves a number of processes, of which the main ones are abrasion, chemical erosion and the hydraulic force of moving water.

*Abrasion*, the erosion of the channel by the impact or grinding action of particles carried by the river, is particularly important. In this way the bed

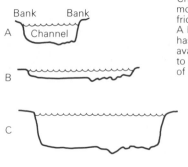

Channel B loses more energy through friction than channel A but channel C has the greatest available energy due to the larger size of the channel.

It is important to note the difference between the river channel and the river valley.

Fig. 5.1 River channel cross sections

and walls of the channel are scoured and excavated. The nature of the load is thus significant in determining a river's erosive power. In the humid tropics active chemical decomposition causes loads to be very fine, reducing the ability of rivers to erode hard outcrops. This factor helps to account for the high proportion of rapids and waterfalls along many African rivers, even in their lower course (page 110).

A special type of abrasion is pothole drilling by coarse sand and pebbles in swirling water, especially where the channel crosses fairly hard rock outcrops. African examples of potholes occur on the Ikiwe River 5 km south of Machakos, Kenya, where it plunges down 150 m through the Marubu Gorge (Plate 5.2); on the Kaduna River 50 km downstream from Kaduna at Badem Pito, Nigeria; and at the confluence of the Blyde and Treur in NE Transvaal, South Africa, known as Bourkes Luck Potholes.

*The hydraulic force* of fast moving water can loosen and wash out weak materials and undermine river banks. It is notably effective in rocks that are well-jointed or well-bedded.

*Chemical erosion* is the solvent action of the water flowing over calcareous rocks like limestone.

In savanna areas, where streams disappear in the dry season, rock weathering plays an important role in the down-wearing of the channel bed.

A river carves its channel by vertical, headward and lateral erosion.

*Vertical erosion* is the deepening of the channel by gradual downcutting, particularly by abrasion.

*Headward erosion*, when the river cuts back at its source and so increases its length, is generally due to undercutting, notably where the ground surface has a protective vegetation cover or there is an upper resistant rock layer. Headward erosion is also concentrated at steep parts of the channel, such as waterfalls, and is especially active when a lowering of base level causes rejuvenation (page 12).

*Lateral erosion*, the wearing away of the channel sides, is most marked at the curved bends or meanders that develop in rivers (Fig. 5.4). As a stream meanders the water reaches its maximum speed against the outside bank of the meander which leads to increased hydraulic force and the consequent undercutting of the bank. Lateral erosion can be very active during floods when the channel width may be enlarged by slumping and caving of the banks.

**River deposition**

Deposition occurs when stream energy is insufficient to carry all the load. Large boulders and other heavy particles are deposited first, followed later by finer debris. Material deposited by a river is given the general name of alluvium. Deposition is greatest in the lower course, especially during floods. The channel is normally too small to hold all the flood discharge, so water spreads out on the adjacent flood plain (page 85). The resulting wide sheet of water encounters increased friction. This, together with the limited gradient, means that speed

Plate 5.2 Potholes in the steep bed of the Ikiwe river where it flows through the Maruba Gorge, south of Machakos, Kenya

is reduced and the river is no longer competent to carry all its load, so deposition takes place.

In the savanna lands the dry season brings a sharp drop in the volume of water leading to extensive deposition, while rivers crossing arid areas suffer especially from large-scale evaporation. The volume of water in the lower Orange River (Fig. 8.2) falls so much that a sandbar is able to block its mouth for nine months a year.

Deposition occurs locally wherever speed is reduced, for example on the inside of meanders (Fig. 5.3), or where there is a sudden decrease in gradient, or where a narrow valley suddenly opens onto a broad plain (Fig. 5.19). In particular, large-scale deposition occurs where a river enters the sea or a lake.

## Characteristic features of the upper course of river valleys

In their upper courses the cross profiles of most river valleys are narrow and V-shaped (Plate 5.3) while the gradient of the valley is usually steep and interrupted by waterfalls and rapids (page 110). Such valleys are most common in upland areas where erosion is more active than deposition. Streams tend to follow a winding course, avoiding resistant outcrops where possible so causing valley spurs to overlap or interlock. The river fills the whole of the valley floor which is constantly deepened by vertical erosion, while the sides are worn back by weathering, mass wasting and down-

Plate 5.3 Narrow V-shaped valleys eroding headwards into the face of the Drakensberg Escarpment, South Africa

Plate 5.4 Interlocking spurs along the narrow upper valley of the Kafue river in central Zambia

wash. In this way the valley attains its V-shape. All these characteristics are apparent in the narrow valley of the upper Kafue River, Zambia (Plate 5.4).

It is in similar narrow valleys that the Nile, the longest river in the world, begins its 6670 km journey to the sea, flowing through 35° of latitude. The Nile headwaters lie in the hills of Rwanda and Burundi between Lakes Kivu and Victoria (Fig. 2.11). Here many small streams carve V-shaped valleys as they make their way north to the Kagera River and Lake Victoria.

In areas of hard rock, valley down cutting may take place faster than weathering and wasting can wear back the valley slopes, with the result that the river flows in a deep, steep-sided valley or gorge (page 113), for example the gorge of the Maletsunyane River in the high Maluti Mountains of Lesotho, with a 190 m waterfall at its head.

On the wide level plains of central Africa many streams have their source in broad, flat-floored, swampy depressions called dambos. Dambos are common in regions of gentle gradient, such as the plateau area of Zambia. Water flowing into the shallow dambos during the rainy season promotes sheetwash (page 72) along their sides and is then held by dense swamp grasses growing along the flat floor. Dambos are widest at their heads, and they narrow downstream as the water flow becomes

more powerful and is able to carve regular valleys. Good examples of dambos occur along the Mulungushi and its tributaries, the Muwofwe and Mteteshi, near Kabwe.

## Landforms characteristic of the lower course of river valleys

In the lower course the cross profile of river valleys is usually broad and flat-floored, while the gradient of the valley is gentle. The valley floor is much wider than the river channel, and the major landform of the lower course is an extensive flood plain, across which the river winds in constantly shifting loops or meanders.

However, not all river valleys have this wide, open cross profile in their lower course. The effects of rejuvenation (page 12) have given narrow, steep-sided profiles to the lower course of many African rivers, such as the Congo and the Zambesi (pages 114–116).

### Meanders

Meanders are the curved bends of a river channel (Plate 5.5). Some streams have straight channels, and some have braided channels (page 85), but these are less common than meandering channels. Straight channels are most apparent in small streams on very steep slopes or where the channel is controlled by joints or faults.

As a stream flows from its upper to its lower course so it gradually widens its valley by vertical and lateral erosion. This process of valley widening is largely the result of stream meandering. As the stream meanders, the river current develops a corkscrew action. The fast-moving water with the greatest energy is directed towards the downstream side of the outer bank causing undercutting, while a return flow of water across the bottom of the channel deposits some of the load against the inner bank (Fig. 5.2). The constant erosion of the concave bank produces a river cliff and the deposition on the convex bank forms a slip-off slope (Fig. 5.3).

During flood times, when discharge is high, the

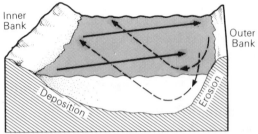

Fig. 5.2 Cross profile of the river channel at a meander bend

fast-flowing current in each meander is displaced downstream so that maximum undercutting takes place at the downstream side of the meander. This process causes the meanders to shift and gradually migrate downstream. The meander belt also moves, swinging back and forth across the valley floor (Fig. 5.4).

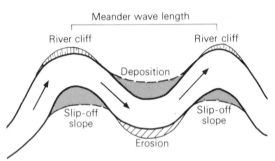

Fig. 5.3 Meander wave length

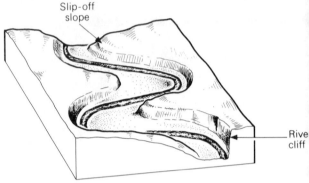

Fig. 5.4 Valley floor widening and flood plain formation through the downstream migration of meanders

Plate 5.5 The meanders of the Ngaila river on the Kano Plains, SE of Kisumu, Kenya. Both river cliffs and slip-off slopes can be seen on the river banks. A cut-off is almost completed on the furthest meander

The origin of meanders is not fully understood, but they are clearly very common, and investigations show them to be related to the mechanism of stream flow. They are initiated by the development of alternate deep and shallow sections that form on the bed of a straight channel and cause the stream to swing from side to side as it is deflected by the shallow sections. Studies show that the width of a channel is related to the discharge at bankfull (i.e. when the channel is full of water to the top of the banks), and also that the wavelength of meanders is 8–12 times the channel width. There is thus seen to be a close relation between river discharge and the size or wavelength of meanders; the greater the discharge, the greater the wavelength.

## Flood plain

The flood plain is a wide flat plain of alluvium on the floor of a river valley across which the river flows in a meandering or braided channel. During floods the river may spread over the entire width of the plain. Most flood plains are marshy, with several swamps and small lakes. The latter are generally the remains of a former channel left behind as the river swings to and fro and migrates downstream. In some flood plains low embankments or levees (page 87) develop on either side of the major channel.

A flood plain develops through the meandering of a river. As the river swings back and forth across the valley, it widens the valley floor (Fig. 5.4). The valley sides and spurs are slowly worn back, and in time a line of low bluffs or river cliffs is formed on either side of a level plain (Fig. 5.5). Eventually this will become so broad that the meanders can swing freely without touching the valley sides. At the same time, the strips of alluvium deposited on the meander slip-off slopes, called point-bar deposits (Fig. 5.6), gradually join up to cover the eroded valley floor with a layer of alluvium, forming an alluvial plain or flood plain.

Over this plain the river may deposit additional sediments during floods, so building a complete cover of alluvium over the bedrock of the valley floor. In its later development the flood plain will be several times wider than the meander belt, but the

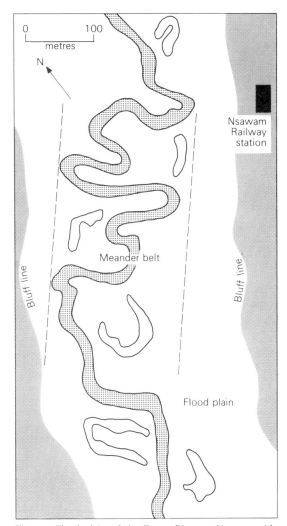

Fig. 5.5 Flood plain of the Densu River at Nsawam with meanders and meander scars

former courses of the river channel are often still visible from the air in the different shades and patterns of the alluvial deposits (Plate 5.6). A particularly extensive zone of marked point-bar deposits, separated by shallow linear swamps, occurs on the south bank of the Chari River in Cameroon, opposite Ndjamena (Fort Lamy).

## Braided channel

A braided channel is an extremely wide, shallow channel in which the river divides and subdivides in

Plate 5.6 The ever-shifting channel of the Zambesi as it flows across the wide, flat Barotse Plain in western Zambia. The point-bar deposits are apparent in the curved strips of alluvium that line the convex sides of meanders

Plate 5.7 The wide braided channel of the Mangoky River, east of Morombe in south Madagascar

a series of interconnecting minor channels separated by sandbanks and islands of alluvium (Plate 5.7). The sandbars and minor channels are generally unstable and during floods may all be submerged. Braiding is most common in heavily loaded streams flowing between banks of easily eroded material. It develops with the deposition of shoals of coarse material, such as sands and gravels, on the river bed. These gradually increase in size and form islands. The erosion of the weak banks widens the channel causing shallowing and increased friction on the bed, which encourages further deposition.

### Oxbows and meander scars

An oxbow is the horseshoe-shaped final section of a once very pronounced meander, now separated from the main stream. Those that still contain water are called oxbow lakes (page 237). Oxbows are formed along parts of the floodplain where meanders are so sharp that only a narrow neck of land remains between two meander loops (Plate 5.5). The narrow neck develops at times of high discharge when maximum erosion of the outer bank is displaced and shifts to the downstream side of the meander curve. Eventually during flood the neck

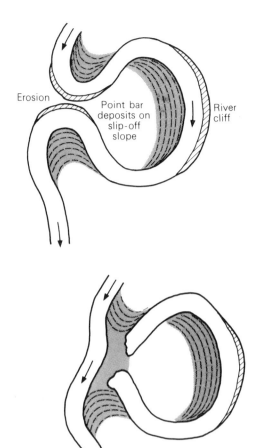

Fig. 5.6 Oxbow lake formation

may be broken through, and the river may bypass the meander, cutting it off by deposition at the old meander neck (Fig. 5.6).

In time the oxbow lake will become filled with alluvium from later floods and will gradually dry out, so that it leaves only a meander scar. Meander scars are often recognisable from their land use or vegetation pattern which contrasts with that over the rest of the flood plain (Plate 5.8).

## Levees and deferred tributaries

Levees are embankments built by some rivers alongside of their channel. They rise above the level of the adjacent flood plain, towards which they slope gradually, and they are wide in relation to their height. Most levees are not more than one or two metres high, but some are much bigger. They encourage lake and swamp formation in the flood plain, since not all water is able to return to the main channel when the floods subside. In densely populated regions these natural formations are often strengthened by man.

Levees are formed through successive flooding and deposition of sediments by the river. Deposition is greatest nearest the river because as the water floods out of the main channel its speed is immediately checked by friction with the banks and the heavier sediments are dropped first. Along rivers where the height of levees has been raised by man to prevent floods there is a tendency for some rivers to deposit along their beds. The eventual effect of this may cause the rivers to flow between their levees at a higher level than their flood plains (Fig. 5.7).

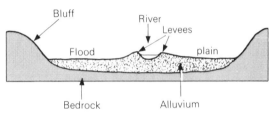

Fig. 5.7 Cross-section of a flood plain

A deferred tributary is a river that is forced to flow alongside the main river for a long distance before being able to join the main river. This deferred confluence is caused by floods building a levee across the original confluence of the two rivers.

### The Lower Sebou River, Morocco

The Sebou flows into the Atlantic north of Rabat (1:100,000 Morocco IGN sheets 5/6, 7/8 Ouezzane). In its lower course it meanders between levees across the flat plain of the Rharb for over 200 km, after descending the steep uplands of the Rif and Middle Atlas Mountains (Fig. 5.8). The load of gravels and sands carried from the uplands is considerable, and on reaching the plain the river begins to deposit on a large scale.

The regime of the Sebou, and its tributaries the

Plate 5.8 One of a number of meander scars in the flood plain of the Wilge River, near Harrismith, South Africa. The swamp grass of the scar contrasts with the surrounding newly ploughed land. The present river channel follows the line of trees in the background

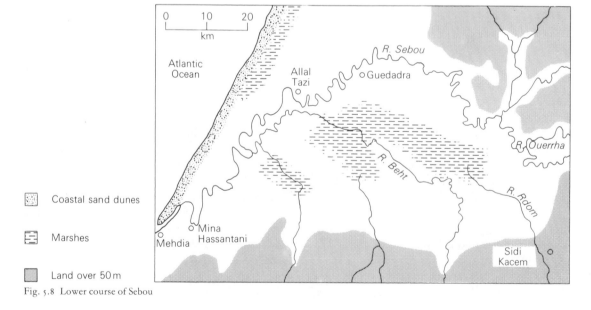

Fig. 5.8 Lower course of Sebou

Ouerrha and the Beht, varies greatly between winter and summer. The winter discharge may be up to 4000 cu metres per sec compared with less than 5 cu metres per sec in August. Whenever the discharge exceeds 2000 cu metres per sec extensive floods occur, since the river, bounded by its levees, is flowing about 3 m above the level of the plains on either side. Serious floods took place in January 1963 when more than 2000 sq km were under water, including much valuable agricultural land.

Along the course of the Sebou and Beht are several oxbows and meander scars, especially upstream from Allal Tazi and near Guedadra. The levees of the Sebou are 200 to 300 metres wide, sloping gently down to the flat plains. Their presence has prevented tributaries joining the main river. Only the Beht is successful, but its confluence is deferred, and it is forced to flow alongside the main river for 20 km. Smaller streams, like the Rdom, end in the large marshes, or merja, that characterise the low-lying parts of the plain.

### The Benue, Nigeria

The Benue River (Plate 5.9) is the chief tributary of the Niger, and for much of its journey between Garoua, in Cameroon, and Lokoja it flows over a wide, flat flood plain on the floor of the Benue Rift (page 20). The river has not developed excessive meanders, but nevertheless the channel is rarely straight and along much of its course it is greatly encumbered by shifting sand banks.

The main floods come between July and September, when the river spreads out over the adjacent clay-covered flood plain. During floods the discharge of the Benue exceeds 10,000 cu metres per sec, while in the dry season this drops to 100 cu

Plate 5.9 The wide channel of the Benue River, near Makurdi, Nigeria

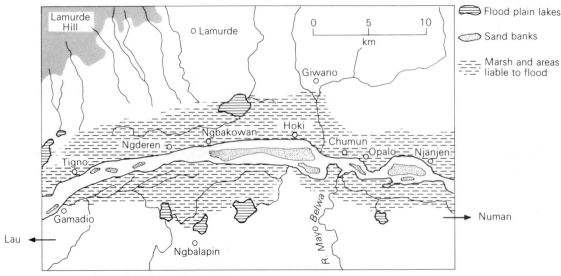

Fig. 5.9 A section of part of the Benue flood plain between Numan and Lau

metres per sec. At certain points the plain narrows as hills close in on either side, for example above Yola and near Makurdi, but mostly it is up to 10 km and more wide (1:100,000 Nigeria sheets 174, 175, 194, 195, 196). In the lower parts of the plain are numerous shallow lakes and swamps connected to each other and the main stream by winding distributaries.

A section of the Benue between Numan and Lau is illustrated in Fig. 5.9. Along this stretch the channel is bordered by levees. The levee on the north side is the most conspicuous and provides the site for many villages, such as Hoki and Ngderen, that are linked by footpaths.

### The braided course of the Congo, below Lisala, Zaire
For much of its long journey the Congo flows through a vast downwarped basin, the central part of which is filled with Tertiary and Quaternary sediments. Between Lisala and its junction with the Kasai the river crosses these recent sediments for nearly 1000 km in a broad braided channel, which in places is over 12 km wide (Fig. 5.10). On either side of the Congo in this section are huge areas of swampland.

Other examples of braided channels occur along the Nile, between Malakal and Khartoum in Sudan; the Orange, between Upington and Augrabies Falls in South Africa and the Kilombero, between Utengule and Ifakara, in Tanzania.

### The Lower Tana River, Kenya
In Kenya only two rivers reach the sea throughout the year, the Tana and the Galana. Along its lower

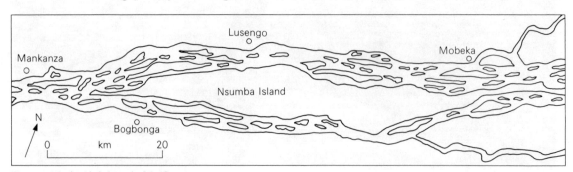

Fig. 5.10 The braided channel of the Congo

course, between Garissa and its mouth, a distance of 250 km, the Tana crosses a low-lying plain in a complicated system of meanders (1:100,000 Kenya sheets 126, 140). For much of its journey the river banks are bordered by thick forest which contrasts with the dry savanna scrub beyond the flood plain. In many sections of the valley the meanders have formed cut-offs and oxbow lakes, especially near Garissa (Plate 11.8) and Nanigi, 50 km south of Garissa (Fig. 5.11).

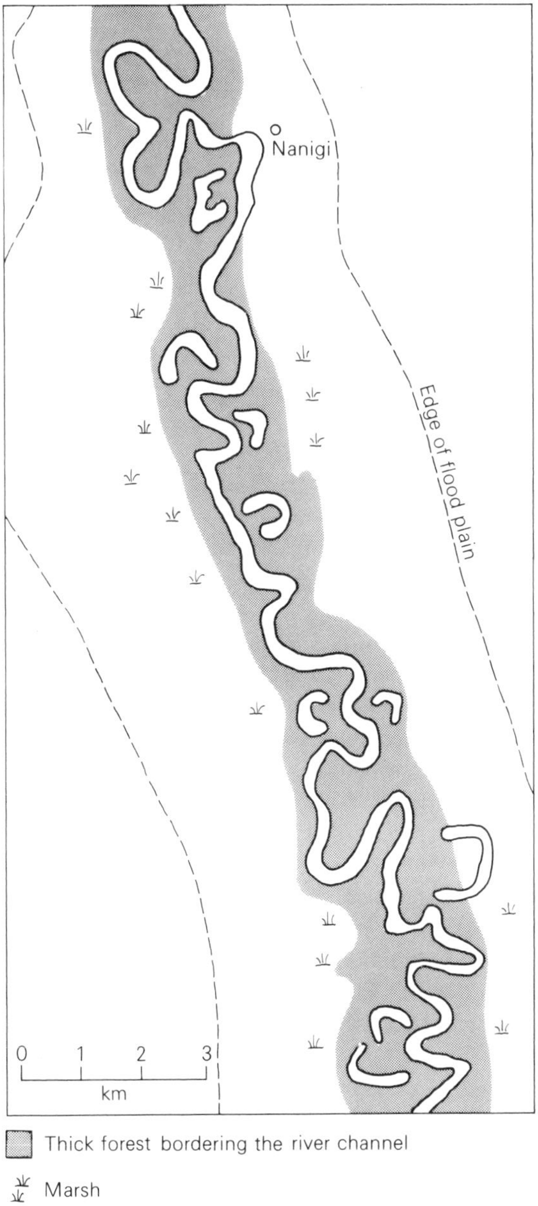

Other examples where characteristics of a river's lower course are common include: the Barotse Plain on the Zambesi, near Mongu, west Zambia (Plate 5.6); the Kilombero, between Utengule and Ifakara, Tanzania; the Lower Sabi, SE Rhodesia (the Sabi has built levees more than a metre in height which has caused several tributaries to have their junctions deferred, for example the Nyantsa and the Msaweze); the Ngaila (Plate 5.5) and Nyando Rivers, flowing into Lake Victoria, south-east of Kisumu, Kenya; the Wilge River, Orange Free State, South Africa, especially near Harrismith, and the lower Densu River, Ghana (Fig. 5.5).

## Deltas and Estuaries

### Deltas

A delta is a large, flat, low-lying plain of river deposits laid down where a river flows into the sea or a lake. Most have a triangular shape and project beyond the shoreline. The delta surface is near sea level and is swampy with several small lakes and lagoons formed by river and marine deposition. Many deltas are covered by swamp vegetation and, in the tropics, mangroves.

Delta formation depends on the amount of sediment deposited relative to the amount removed by waves and tidal currents. Rivers with a small load flowing into oceans where both wave energy and the tidal range are high are unlikely to form deltas. Similarly, rivers with a large sediment load (Fig. 5.17) but a high stream velocity, like the Congo, may not form deltas since the river itself is able to wash the sediment far out to sea. But rivers with a relatively large load and low velocity near the mouth, like the Niger, tend to form deltas even in major oceans.

The shape and size of a delta is determined by the nature and quantity of sediment, together with the strength of waves and tidal currents. Deposition results both from the reduction in speed as the river enters the sea and also from coagulation of fine

Thick forest bordering the river channel

Marsh

Fig. 5.11 Cut-offs and oxbow lakes near Nanigi on the Tana River, Eastern Kenya

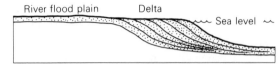

Fig. 5.12 Diagrammatic section through a delta

material mixing with salt water. Initially material is splayed out on the continental shelf around the river mouth, but in time it appears above sea level as new land (Fig. 5.12). Deposited sediments often block the existing river channels causing new ones to form, and this, with the growth of vegetation, encourages further deposition, so increasing the size of the delta.

There are three main types of delta:

1  Arcuate – built by rivers with many distributaries carrying both fine and coarse materials, and where offshore currents are strong enough to round the delta's seaward edge.
2  Estuarine – built by rivers depositing materials in a submerged river mouth forming sand banks and islands, around which wind several distributaries.
3  Birdfoot – formed by rivers carrying large loads of mainly fine material into water where wave energy is low. A few very long distributaries, bordered by levees, jut out from the shore.

## Estuaries

An estuary is a broad tidal channel at the mouth of a river where it enters the sea. Some estuaries are deep and fairly narrow, and deposited sediments are soon removed by tidal scour. Other estuaries are shallow and low-lying, and deposition is slowly filling them in. Most present estuaries resulted from submergence due to a recent rise in sea level (see page 13).

### Niger Delta

The Niger delta is the largest in Africa, covering more than 35,000 sq km. It is arcuate in type with a convex front forming a broad arc of about 350 km from the Benin to the Bonny River (Fig. 5.13). The delta is crossed by numerous distributaries bordered by levees and separated by areas of lower lying land. Altogether there are nearly 20 mouths

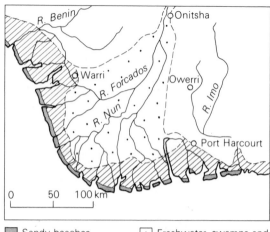

Sandy beaches  •  Freshwater swamps and forests

Mangrove  ▢ Older rocks

Fig. 5.13 The Niger Delta

along the delta front, but only two main ones, the Nun and the Forcados. High rainfall totals and annual floods cause the distributaries to change course and leave abandoned channels and small lakes throughout the delta area.

The outer delta is fronted by a zone of sandy beaches backed by mangrove swamps and creeks (page 216). Inland, towards the head of the delta, the vegetation changes to freshwater swamps and dense forest.

The Niger has been building this delta since the Cretaceous period and the break-up of South America from Africa (page 10). But a large part of the sedimentary deposits are very recent in age, having been laid down only in the last 12,000 years. The total depth of deposits in the delta is thought to exceed 10,000 m. This enormous accumulation has been possible because the crust has been isostatically depressed by the great weight of the sediments (see page 8).

Other examples of arcuate deltas are: the River Nile, Egypt (Plate 5.10); the River Mangoky, Madagascar, and the River Tsiribihina, Madagascar.

### Betsiboka Delta, North West Madagascar

The Betsiboka delta is one of several on the west coast of Madagascar, but in contrast to the other arcuate types this is an estuarine delta. Although fairly large, the Betsiboka delta is small when compared to the Niger. Almost the entire area of the

Plate 5.10 The Nile Delta, Egypt, bow-shaped and bordered on the coast by lagoons and sandspits. The completion of the Aswan High Dam and formation of Lake Nasser has reduced the Nile's load. The silt-free water now flowing past Cairo has caused river bed erosion and limited the rate of delta growth, while along the coast erosion by west to east currents has now become so active that some parts of the delta are receding by several metres a year

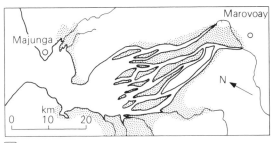

Mangrove

Fig. 5.14 The Betsiboka Delta

delta is covered by mangrove, but, as is typical of such deltas, its area is limited to the form of the drowned river mouth that is gradually being infilled (Fig. 5.14). Much of the material for the infilling is debris washed down from the lavakas (page 76) that gash the slopes of the central highlands.

*Medjerda Delta, Tunisia*
The Medjerda River has built up the greater part of its delta during the last 5000 years, by gradually filling in an arm of the sea off the Gulf of Tunis (Fig.

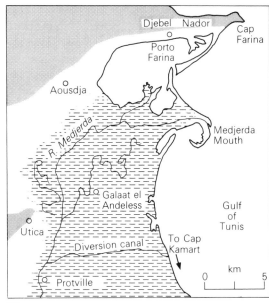

Land over 50 m

Land liable to flood

Fig. 5.15 The Medjerda Delta, Tunisia

5.15). The town of Djedeida, now 40 km from the Medjerda mouth, once lay on the Mediterranean coast. In the 8th century BC the coast extended from Cap Farina (Ras Sidi Ali el Mekki) to Cap Kamart, by way of Utica and Galaat el Andeless. Utica, founded about 1100 BC, was a major port in the ancient world and remained so until about the 4th century AD. Then, following the felling of forests and consequent extensive erosion of hillsides, the sediment carried by the Medjerda increased rapidly and it began depositing on a very large scale. The result was that by 1900 Utica lay 10 km inland.

The construction of the delta, which now covers about 750 sq km, was greatly assisted by the position of Cap Farina which deflected the Mediterranean currents away from the river mouth and thus reduced the rate of sediment removal. The present mouth of the Medjerda is now nearly 20 km north of the position it occupied in Carthaginian times. Estimates show that to build such a delta the average annual load carried by the river must have exceeded 15 million tons. Deposition was particularly great around 1950, and the delta front, advancing rapidly, was curved by wave and current

93

action into a crescent shape (1:200,000 Tunisia IGN, Bizerte sheet). However, recent engineering works, such as dams and diversion canals, together with afforestation and soil conservation, have reduced the load by over half and the delta is no longer enlarging itself at the former rate. The most significant of these works is the diversion canal which leads off water north of Protville. The overall effect is that the crescent-shaped delta mouth is fast being destroyed by marine erosion (see also page 212).

## Omo Delta, Ethiopia

At the north end of Lake Turkana (Rudolf), on the Ethiopia–Kenya frontier, the south-flowing Omo River (Fig. 2.4) has formed a birdfoot delta. The waters of Lake Turkana (Rudolf) are very dense due to a high soda content, making it possible for the Omo to carry its load several kilometres out beyond the lake shore. In depositing the sediments the river has built up a system of parallel levees either side of three main distributaries (Fig. 5.16).

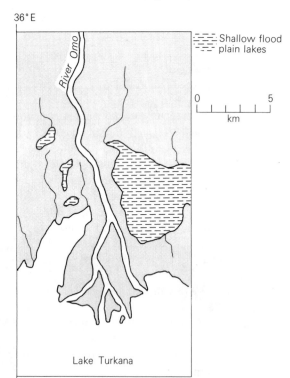

Fig. 5.16 Birdfoot delta of Omo River, Ethiopia

## Inland deltas in Africa

Two of the largest inland deltas in the world are the Niger in Mali and the Okavango in Botswana. They are areas of extensive deposition, where the rivers subdivide into numerous distributaries and flow between levees, flooding annually over vast areas of swampland. Both the Okavango (Plate 5.11 and Fig. 8.2) and the Niger inland deltas are the remnants of former lakes (page 241).

## Congo Estuary

The Congo is the only one of Africa's major rivers with a large deep water estuary (1:50,000 British Admiralty Chart 638). This is mainly the result of the high water discharge relative to the sediment discharge (Fig. 5.17), together with the great energy of the river. Near Matadi speeds of 500 cu metres per sec have been recorded.

| | Annual water discharge (million cubic metres) | Annual sediment discharge (million tons) |
|---|---|---|
| Congo Niger Nile | 1,400,000 180,000 85,000 | 70 40 60 |

Fig. 5.17 High water discharge relative to the sediment discharge

The estuary begins at Boma (Fig. 5.18) and extends for about 80 km to the Atlantic Ocean. It is 10 km wide at the mouth. The upper part of the estuary needs regular dredging due to innumerable sandbanks and islands, while at the mouth, below Malela, there are powerful currents that scour away deposited materials.

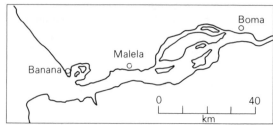

Fig. 5.18 The Congo Estuary

Plate 5.11 Part of the vast Okavango inland delta in northern Botswana

Below the surface of the Atlantic the estuary is prolonged by a gigantic V-shaped submarine canyon. It is one of the largest in the world and cuts down the continental shelf onto the ocean floor for over 700 km depositing a huge debris cone at its lower end. Its depth is over 1000 m in places, and this is thought to be due to the strong underwater turbidity currents that carry great quantities of sand and mud to the ocean floor.

Other examples of estuaries occur on the Gambia River (Plate 10.13); the Gabon River, Gabon; the Cross River, Nigeria; and the Sierra Leone River (Fig. 3.20).

## Alluvial fans

An alluvial fan (debris fan) is a fan-shaped deposit of fairly coarse material laid down by a stream with a large load as it emerges from a steep, narrow valley, such as a ravine or gully, onto a wide gentle plain. A sharp increase in channel width and decrease in gradient causes a marked reduction in stream energy resulting in sudden deposition (Fig. 5.19). Fans range in size from a few metres to several kilometres. They are similar to deltas, except that they

are formed on land and are therefore sometimes called 'dry deltas'. They tend to have a steeper gradient than deltas, sloping up towards the apex. The river forming the fan usually flows across it in a system of braided channels. A fast-growing fan developing from a tributary flowing into the main valley may block the main river and form a lake (see page 236). Many alluvial fans are active and still growing, while others are now clothed in vegetation and no longer active.

In some areas where a steep mountain front abuts onto a lowland a series of coalescing alluvial fans

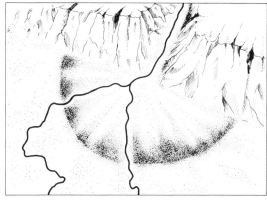

Fig. 5.19 Diagram of an alluvial fan

may develop and form a continuous piedmont alluvial plain at the mountain foot. Fans built of very coarse material and with a steep slope are called alluvial cones or debris cones (page 166).

*Alluvial fans in Tanzania*

In south Tanzania, the broad Kilombero Valley, through which the Kilombero River follows a meandering and braided course, is bordered by the steep-sided highlands of Uchungwe on the north-west and Mahenge on the south-east. Along both sides of the valley, at the foot of the highlands, alluvial fans have formed where streams leave constricted valleys and flow out onto the flat flood plain (1:50,000 Tanzania sheet 235/1). There are more than 15 fans, built largely of clays and sands, between Utengule and Ifakara, for example the Rondo, the Rufiri, the Lumemo and the Luri (Fig. 5.20).

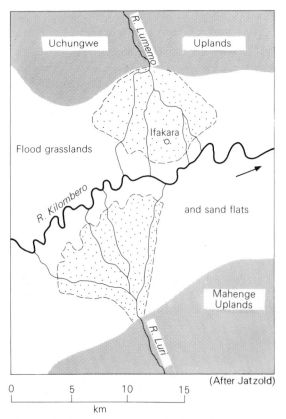

Fig. 5.20 The Luri and Lumemo alluvial fans in the Kilombero Valley, Tanzania

Other examples are:

1 Lume fan: formed by the Lume River as it reaches the Semliki Plain on the west of Ruwenzori (Fig. 2.11). The fan has even forced the Semliki to take a more westerly course.
2 Blida fan: Blida (40 km SW of Algiers), Algeria, lies at the apex of a large old fan formed by an ancient stream where it left the foothills of the Tell Atlas Mountains. The fan has a radius of over 5 km and is one of many forming the large Mitidja Plain south of Algiers (1:25,000 Algeria IGN sheet Blida 63 1–2).
3 Amekrane fan: one of several on the east of the Nekor Valley, SE of Al Hoceima, Morocco (1:50,000 Morocco IGN sheet N1-30-XXI-1a). (See also page 236)

## Landforms caused by rejuvenation in a river valley

Rejuvenation refers to renewed erosive activity in a river valley and usually results from a negative movement of base level, caused either by a fall in sea level or a regional uplift of the land (page 12).

A river may also be rejuvenated by an increase in its discharge causing increased energy. This may be due to increased rainfall or perhaps river capture (page 101).

A positive movement of base level produces the opposite effects. The lower course of river valleys are drowned and the rivers begin extensive deposition. The characteristic landforms are considered in Chapter 10, as part of submerged coastlines (page 217–18).

### River terrace

A river terrace is a step or bench cut in the side of a river valley and covered by a layer of gravel and other alluvial deposits. When base level falls and a river renews its erosive activity, some parts of the former flood plain and underlying rock floor may not be removed by lateral erosion and will remain on the valley side as terraces (Plate 5.12). These rejuvenation terraces are generally of equal height

Plate 5.12 River terraces in the lower part of the Sani Pass, NW of Himeville, on the Lesotho–South
Africa border

on either side and are called paired terraces. If
several falls in base level have taken place, a
succession of terraces will result, separated from
each other by a marked break of slope. Some
terraces are quite wide and extend along a valley for
many kilometres, while due to greater lateral
erosion others are small and discontinuous.

Terraces may also form from the downstream
migration of meanders (see page 84). Meander

terraces are generally small, unpaired and cut in
alluvial deposits. They are not related to any change
in base level, but result from incision by the river
caused by increased energy, perhaps due to higher
rainfall or a reduced load.

*River terraces along the Upper Birim River, Ghana*
Two major terrace levels may be identified along the
Birim Valley, near Kibi, a Lower Terrace and a

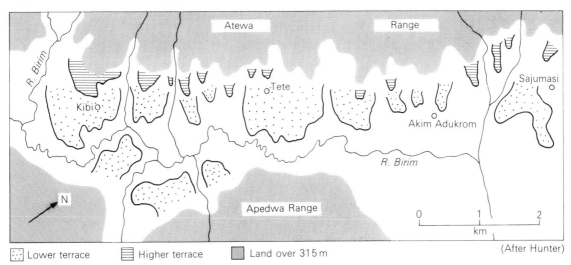

Lower terrace    Higher terrace    Land over 315 m    (After Hunter)

Fig. 5.21  River terrace remnants along the Upper Birim, Ghana

Higher Terrace (1:62,500 Ghana sheet 94). Between Kibi and Sajumasi the Lower Terrace is about 270–300 m above sea level, while the Higher Terrace is about 300–315 m above sea level. The remnants of the younger, Lower Terrace are far more complete than the Higher Terrace and are covered by coarse well-rounded gravel deposits. The height of the terrace above the river increases downstream from Kibi as the river becomes more incised (Fig. 5.21).

Near Kibi the Birim has cut into its valley to a depth of about 2 m below the Lower Terrace. Within a distance of less than 10 km downstream the depth of incision exceeds 20 m. Between Kibi and Tete knickpoints have been located along the river which can be correlated with the Lower Terrace levels. At Akim Adukrom (Chiripon) the terrace is nearly 15 m above the river, and the new 1970 road alignment west of the village clearly reveals old river gravels in the cuttings.

Similar terrace levels of the Birim also exist further downstream, especially near Oda and Kade.

Other examples of rejuvenation terraces occur on the Oti River, near Demon and Saboba, north Ghana, at about 14 m, 20 m, 27 m and 33 m above the river bed; the Niger River, near Onitsha, south Nigeria, at about 10 m, 30 m and 80 m above high water level; and on the Vaal River, near Warrenton, South Africa, at about 21 m, 30 m and 60 m above the river bed.

Meander terraces have been recorded at heights up to 10 m above the bed of the Chongwe and Chalimbana Rivers east of Lusaka, Zambia, and also along some of the rivers flowing into Lake Victoria, SE of Kisumu, Kenya, for example the Nyando and the Ngaila (Plate 5.5).

## Incised meander

An incised meander is the curved bend of a river that has been incised into the land's surface so that the river now winds between steep valley walls. Most incised meanders result from the rejuvenation of an already meandering stream, due to a fall in base level. There are two main types:

1 *ingrown* – with an asymmetrical valley cross profile, that is alternate steep sides with undercut

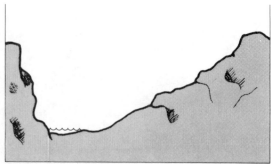

Fig. 5.22 Ingrown incised meander

slopes facing alternate gentle sides with large slip-off slopes. Ingrown meanders develop on resistant rocks and where base level falls gradually, hence meanders tend to shift laterally as they are incised producing the asymmetrical profile (Fig. 5.22);

2 *intrenched* – with a steep-sided symmetrical cross profile. This type develops on weak rocks and where base level falls quickly, causing rapid vertical incision.

Incised meanders also develop meander cut-offs similar to those found in flood plains, except the abandoned meanders are incised. However, it should be noted that not all incised meanders are due to rejuvenation. Some may originate from straight streams deepening their valleys while at the same time forming meanders.

## Valley within a valley

Along rivers where rejuvenation was fairly rapid and the fall in base level quite large the effect may be to produce a steep-sided rejuvenation gorge within the former valley (Fig. 5.23).

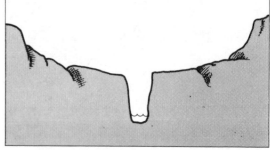

Fig. 5.23 Valley within a valley

Plate 5.13 An incised meander of the Mwachi River, west of Mazeras, Kenya. View looking south from point A in Fig. 5.24

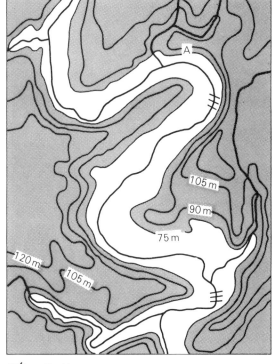

 Rapids

0               1
‎      km

Fig. 5.24 Incised meanders of the Mwachi River, 3 km east of Mazeras, Kenya

## Incised meanders in Kenya and South Africa

Fifteen kilometres NW of Mombasa, near Mazeras, the Mwachi River and its neighbours the Kombeni and Cha Shimba have incised meanders cut into the Mazeras sandstones. The Mwachi meanders are ingrown, and incised to a depth of nearly 100 m (Plate 5.13). On the outside bends of the meanders the valley sides are very steep and often form vertical cliffs in the Triassic sandstones (1:50,000 Kenya sheet 198/3). In the river bed a number of rapids suggest the existence of knickpoints working their way upstream (Fig. 5.24).

In South Africa, between East London and Durban, there are more than ten large rivers along which incised meanders, both ingrown and in-trenched, have developed on a very extreme scale. Among the more important of these rivers are the Umgeni (Mgeni), Umzimkulu (Mzimkulu), Um-tamvuna (Mtamvuna) and Bashee (Mbashe). On one part of the Bashee, 50 km south of Umtata, known as 'The Collywobbles', the meanders are incised about 450 m into the land surface and are so extreme that only narrow 'swansnecks' of land remain between the meander curves (1:50,000 South Africa sheets 3128 DC, 3228 BA).

Incised meanders also occur along the Sine River, an upper tributary of the Mkomazi between Mg-

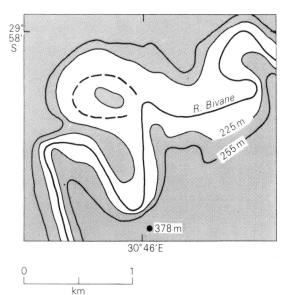

Fig. 5.25 An abandoned incised meander, 20 km SW of Durban, South Africa

washi and Kifungilo in the Usambara Mountains north of Lushoto, Tanzania (1:50,000 Tanzania sheets 109/2, 109/4).

A good example of a rejuvenation gorge within a former valley is the 135 m deep Storms River Valley (Plate 5.14), 50 km east of Plettenberg Bay, South Africa. The Storms River has incised itself, along a distance of 10 km, into the vertically dipping Malmesbury Beds that outcrop between the Tsitsikamma Hills and the sea.

## Knickpoint

A knickpoint is a break of slope in the long profile of a river valley, the long profile being the curved profile that joins a river's source to its mouth. If base level falls, a break will appear in the slope of the long profile which will gradually work its way upstream. The knickpoint is the point in the river

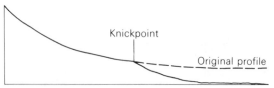

Fig. 5.26 Long profile of a river after rejuvenation

bed where the new profile changes to the old (Fig. 5.26), and its position is sometimes marked by a waterfall or rapid. A river may have several knickpoints and some may persist in one position for a long time if they are held by a band of hard rock.

### Knickpoint at Charlotte Falls, Sierra Leone

At Charlotte Falls, south of Freetown, the Orugu River flows over a series of rapids before continuing SE to Charlotte and Mortema (Fig. 5.27). The rapids, at about 180 m above sea level, represent a retreating knickpoint, above and below which the river's long profile shows only slight irregularities

Plate 5.14 Storms River Valley, South Africa. A rejuvenation gorge incised within the former valley, the floor of which is level with the foot of the Tsitsikamma Hills in the background

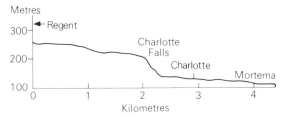

Fig. 5.27 Charlotte Falls knickpoint

(1:10,000 Sierra Leone sheets 8 & 9). Another rejuvenation point can also be identified at Bathurst Falls on the Kongo River, an upper tributary of the Orugu, at about 330 m above sea level.

Other examples of knickpoints occur on the Vaal River, South Africa, below Barkly West, at about 1050 m above sea level; and on the Birim River, Ghana, below Kibi, at about 292 m above sea level (Fig. 5.21).

## River capture

River capture or river piracy is the diversion of part of a river course into the system of an adjacent powerful river able to erode its valley more rapidly than its weaker neighbour. It includes both small-scale captures of single streams and also captures of entire river basins.

For capture to be successful the capturing stream will be flowing at a lower level than its victim, and have greater energy for vertical and headward erosion. This may be due to either:

a) the pirate river flowing over easily eroded rocks, such as weak rocks or rocks that have been broken by faulting (page 25); or

b) the pirate river flowing down a much steeper gradient than its victim, for example the scarp of a cuesta (page 118) or plateau. A stream flowing down a gentle back slope may be captured by a powerful scarp stream cutting back across the watershed. This process is called divide migration or watershed retreat.

Possible evidence of river capture (Fig. 5.28) may include:

1 An elbow of capture, or sharp change in the direction of a river course at the point of capture. But not all right-angled bends in rivers are due to capture, many are a result of structural control (page 105).

2 A wind gap: the valley of the beheaded stream below the point of capture may be left as a dry valley or wind gap. The floor of this former valley may be lined with old river gravels and other alluvium. In time, rivers in the area will be flowing at a lower level and the dry valley will only appear as a gap or col in the adjacent hills.

3 A misfit stream. The beheaded stream having lost its headwaters may be reduced in volume, causing it to appear too small for its valley. It is therefore described as a misfit or underfit stream.

4 Incision of the pirate stream near the point of capture. Due to increased erosive power from its enlarged headwaters the pirate stream may be rejuvenated and incise its valley below the point of capture. If the pirate stream was flowing at a much lower level than its victim, the capture will cause a fall in base level for the captured stream, and thus produce a knickpoint waterfall that slowly retreats upstream.

### Tiva River capture, Kenya

In east Kenya the lower Tiva River has captured a former tributary of the Galana (1:250,000 Kenya sheet SA-37-10). The upper Tiva used to flow south along the east foot of the Yatta Plateau to join the Galana through a gap near Mopea, but the lower Tiva effected a capture north of Wathuni (Fig. 5.29). The evidence of capture is apparent in the elbow bend of the Tiva near Wathuni, and especially in the former valley trace between Wathuni and Mopea. Although this original valley is now at a higher level

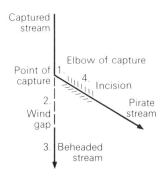

Fig. 5.28 Features of river capture

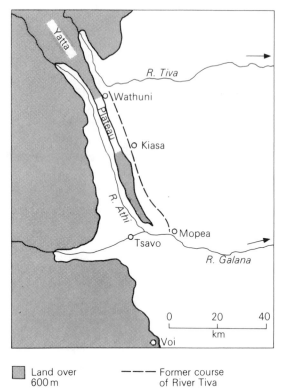

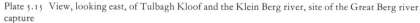

Land over 600 m

- - - - Former course of River Tiva

Fig. 5.29 Location of the Tiva River capture

than the present Tiva it contains thick deposits of former river alluvium, and along many sections near Kiasa swamps still occur.

*Great Berg River capture, South Africa*
In the SW Cape the NW-flowing Great Berg River has captured the headwaters of the SE-flowing Bree (Breede) River in the area of Wolseley (Figs. 2.22 and 5.30). The streams near Tulbagh used to flow SE to the Bree, but the Klein Berg, a tributary of the Great Berg, cut back along the north part of the 500 km Worcester fault, where the Table Mountain Sandstone rocks are shattered, and diverted these streams to the NW (1:50,000 South Africa 3319 AC). The narrow pass through which the capture took place is called Tulbagh Kloof (Plate 5.15).

*Cunene River capture, Angola*
The upper Cunene used to flow south from the Bihé Plateau to a large inland drainage basin, similar to the Okavango Delta (Plate 5.11), but it was captured by a vigorous coastal stream and now turns west to the Atlantic (World Aeronautical Chart 1:1,000,000 sheet 3179). The former inland delta area is now occupied by the Etosha Pan.

Plate 5.15 View, looking east, of Tulbagh Kloof and the Klein Berg river, site of the Great Berg river capture

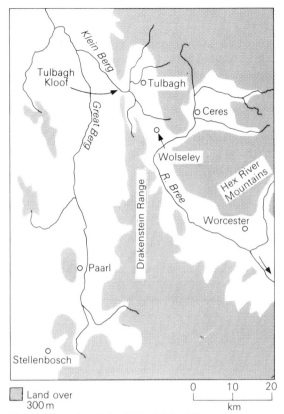

Land over 300 m

0    10    20
     km

Fig. 5.30 Location of the Tulbagh Kloof River capture near Cape Town, South Africa

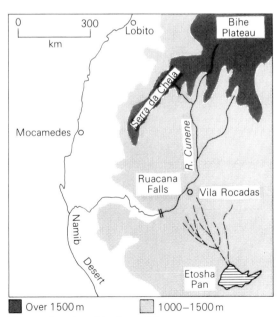

Over 1500 m     1000–1500 m

Fig. 5.31 Location of the Cunene River capture

For the first 600 km of its course the river flows in a wide alluvial plain over the plateau of southern Angola, but 50 km below Vila Rocadas it begins to plunge down the escarpment towards the coast in a series of falls and rapids, notably the 124 m high Ruacana Falls (Fig. 5.31). In this lower section the Cunene descends 1000 m in a distance of about 350 km.

The capture was achieved by the powerful lower Cunene, with its steep gradient and much lower base level, actively eroding back into the plateau edge. The plateau has been slightly downwarped and this was probably sufficient to encourage the upper Cunene and some of the former Etosha Lake to overspill and take its new course. Seasonal water courses still flow south to the Etosha Pan along the former line of the Cunene.

### Nsaki River capture, Ghana

In the Akwapim Hills west of Aburi the Nsaki, a tributary of the Densu, has captured the Bisiasi (1:62,500 Ghana sheet 62). The Bisiasi formerly flowed into the Niensi, an upper tributary of the Volta, to which it is still parallel, until it was diverted into the Nsaki near the village of Bisiasi

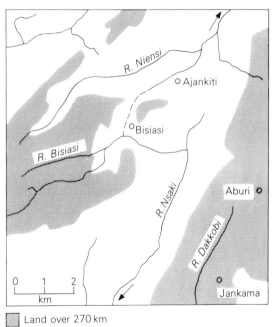

Land over 270 km

Fig. 5.32 Location of the Nsaki River capture, near Aburi, Ghana

(Fig. 5.32). The former course of the Bisiasi is shown in Fig. 5.32 by a dotted line.

The evidence for the capture is three-fold: a wind gap in the form of a col at 250 m north of Bisiasi village, an elbow of capture south of the village, and a deeply incised valley immediately below the point of capture.

Other examples are: the capture of the Enyong, a tributary of the Cross, by the Imo cutting back along a zone of faulted rocks, in SE Nigeria (1:250,000 sheet 79); the capture of the Khubedu, a tributary of the Orange, by the Tugela eroding headwards into the steep face of the Drakensberg scarp, South Africa; and the capture of the Chire, an upper tributary of the Luangwa, by the North Rukuru cutting back into the granitic Nyika Plateau along a zone of weaker rocks, in north Malawi.

## Drainage patterns

A drainage pattern is the layout or plan made by rivers and their tributaries on the landscape. The main types of pattern are differentiated according to their relationship with:
1 the slope of the land;
2 differences in rock hardness;
3 the rock structure.

### Dendritic pattern
It is shaped like the trunk and branches of a tree (Fig. 5.33), the tributaries converging on the main

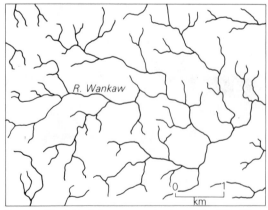

Fig. 5.33 Dendritic drainage on granite about 30 km NE of Cape Coast, Ghana

stream from many directions and usually joining at acute angles. It is the most common type and is not related to rock structure or differences in rock hardness, but tends to form on massive crystalline rocks, like granite, or horizontal to gently dipping sedimentary strata.

### Trellis pattern
A rectilinear pattern in the shape of a lattice with the chief tributaries joining the main stream approximately at right angles (Fig. 5.34). In turn, minor tributaries join the chief tributaries at right angles and flow more or less parallel to the main stream. The pattern is strongly related to structure or differences in rock hardness, and is commonly found in scarpland areas and regions of folded rocks. The chief tributaries are usually aligned along downfolds or parallel zones of weak rock separated by resistant uplands. The only parts of the pattern that are usually not related to structure are the main streams which may cut through the folded uplands, sometimes along a fault or other line of weakness (see also Fig. 6.5).

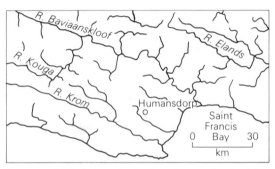

Fig. 5.34 Trellis drainage in the Cape Ranges, South Africa. Rivers follow the weak Bokkeveld shales, leaving the hard Table Mountain sandstone as ridges

### Radial pattern
This is an arrangement of streams flowing outwards down the flanks of a dome or cone-shaped upland, such as a large volcano (Fig. 5.35). It is controlled by the slope of the land.

### Rectangular pattern
A pattern similar in plan to the trellis, with tributaries joining each other at right angles. But the rectangular also tends to have individual streams taking sharp angular bends along their course. It is

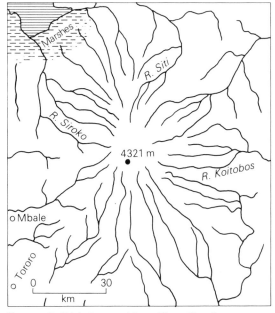

Fig. 5.35 Radial drainage on Mount Elgon, Uganda

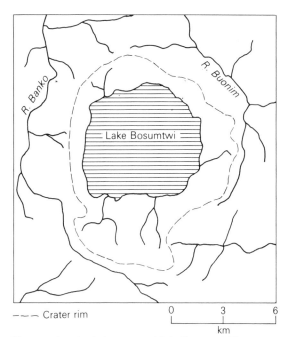

Fig. 5.37 Annular drainage around Lake Bosumtwi, Ghana

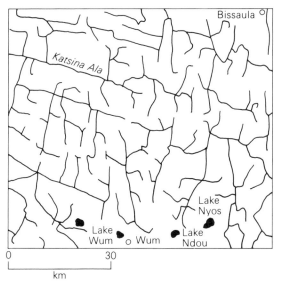

Fig. 5.36 Rectangular drainage and crater lakes in the Bamenda Highlands, Cameroon

the result of structural control, with streams following joints or fault lines in the rock (page 25 and Fig. 5.36).

*Annular pattern*

A pattern with streams often joining at sharp angles, but arranged in a series of curves about a dissected

dome, basin, or crater area (Fig. 5.37). On a dissected dome with alternating bands of hard and soft rock the pattern may appear as several concentric curves.

## Accordant and discordant drainage

In many areas drainage systems and individual streams often show a strong relationship to the geological structure. Streams naturally flow along those lines where they find erosion easy, and for this reason they follow areas of weak rock in preference to resistant outcrops. Fault-guided valleys (page 25) are common in several regions, while fault lines and joints may be reflected in drainage patterns (Fig. 5.36).

In zones of folded rock the initial valleys will be determined by slope and will tend to follow the synclines. Later, due to variations in rock hardness, erosion may carve valleys along the anticlines (Fig. 5.38) and leave the synclines as uplands. Thus it is possible to have both synclinal and anticlinal valleys (Fig. 5.39 and page 31). In instances where drainage is related to structure the drainage is described as

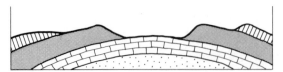

Fig. 5.38 Diagram of an anticlinal valley with infacing scarps

being accordant. Drainage systems that are opposed to the dominant structure are known as discordant or inconsequent.

In Nigeria the Bolleri River and its tributaries (Fig. 5.39) occupy an anticlinal valley in the folded sediments of the Benue Rift (page 20) between Bambuka and Bambam, north of Lau. The Bima sandstone along the crest of the Lamurde anticline has been eroded away to produce a breached anticline with infacing scarps.

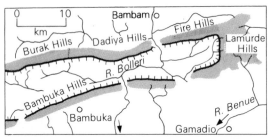

Fig. 5.39 The anticlinal valley of the Bolleri River, Nigeria

There are two main types of discordant drainage – superimposed and antecedent. In both cases river valleys, and sometimes whole drainage systems, are transverse to the geological structure.

## Superimposed drainage

Superimposed drainage describes a river valley or part of a valley that developed on a former cover of rocks and which is now superimposed onto a previously buried and completely different rock structure to which it is discordant.

Superimposed or superposed river valleys are younger in age than the structures which they cut across. They develop in areas where an ancient land surface once covered by sedimentary rocks or lava flows, for example, is now exhumed by the erosion of the rock cover. Streams that originated on the rock cover may be able to carve their valleys down through it onto the buried structure (Fig. 5.40). The

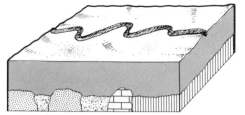

Before removal of cover rocks

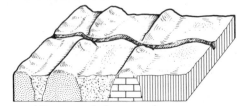

After removal of cover rocks

Fig. 5.40 Superimposed drainage

evidence for superimposed drainage is most obvious when remnants of the former rock cover are still to be seen on either side of the river valley.

## Antecedent drainage

Antecedent drainage describes a river valley or part of a valley that developed on a former landscape later uplifted by local earth movements, but which was able to maintain its course by eroding vertically at a rate that was fast enough to keep pace with the rising land.

Antecedent river valleys are therefore older in age than the structures to which they are discordant. They are most common in areas of fairly recent earth movements, such as the young fold mountains, where uplift took place slowly enough for a river to maintain its downcutting, and where there has been insufficient time for the drainage to become completely adjusted to the structure (Fig. 5.41).

As uplift begins it may have the effect of reducing the gradient of the river bed upstream of the rising barrier, which may cause deposition. On the downstream side the gradient may be increased leading to incision. However, if uplift is too rapid or the river is not powerful enough to develop an antecedent course the drainage may be reversed and lakes may be ponded up (page 24).

The proof of antecedence is often difficult, but

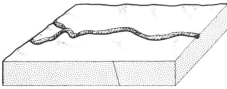

Before uplift

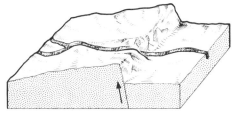

After uplift by faulting

Fig. 5.41 Antecedent drainage

possible evidence that might exist includes deposited sediments upstream of the discordant structure and marked incision downstream.

The essential difference, therefore, between antecedent and superimposed drainage is that the former is older than the structure it crosses, while the latter is younger. One of the most obvious signs that a river may be antecedent or superimposed is when it follows a discordant course through an upland by means of a deep gorge, on either side of which are extensive lowlands.

*Superimposed gorge of the Nile at Sabaloka, Sudan*

After flowing across the clay plain of central Sudan the Nile cuts through the Sabaloka Hills in a narrow gorge, some 80 km north of Khartoum (Plate 5.16).

Plate 5.16 View SE from near El Hugna of the Nile superimposed in a deep gorge onto the igneous ring complex of the Sabaloka Hills, Sudan

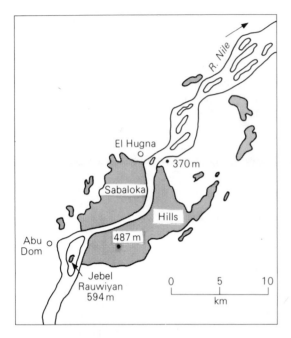

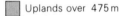
Uplands over 475 m

Fig. 5.42 The River Nile superimposed at Sabaloka, Sudan

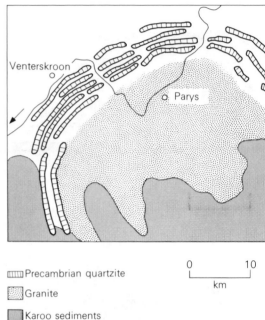

⊞ Precambrian quartzite

▢ Granite

▨ Karoo sediments

Fig. 5.43 The Vaal superimposed near Parys, South Africa

The flat-topped hills are formed of a huge ring complex, consisting of rhyolite lavas partly surrounded by a granitic ring dyke (page 53). The superimposition of the river originated from a former cover of Nubian sandstone rocks, since removed by erosion. The 100 m deep gorge carved by the Nile was a result of the great hardness of the underlying igneous rocks. Evidence of the former cover of sandstone rocks is to be seen in occasional residual hills overlying the igneous rocks, notably Jebel Rauwiyan (Fig. 5.42). Finally, although the superimposition itself is clearly an instance of discordant drainage, it should be noted that many small streams and gullies crossing the Sabaloka Hills show a close adjustment to the main joint pattern of the rock (1:250,000 Sudan sheet NE-36-N).

### The Vaal River superimposed on the Vredefort Dome, South Africa

One hundred kilometres SW of Johannesburg, near Parys, the Vaal has been superimposed onto the formerly buried quartzite ridges of the Precambrian Vredefort Dome, from a cover of Karoo sedimen-tary rocks (Fig. 5.43). The Vaal follows a discordant winding course in this region, twice flowing across the northern part of the circle of ridges, and onto the ancient granite rocks of the core, first about 10 km upstream of Parys and second between Parys and Venterskroon (1:50,000 South Africa sheets 2627 CD, 2627 DC). The major evidence for the superimposition, besides the obvious discordance, is the extensive cover of Ecca sandstones and shales of the Karoo period that still conceal the southern part of the dome.

### The Great Ruaha River, antecedent in the Iringa Highlands, Tanzania

The Great Ruaha begins its course in SW Tanzania and for the first part of the journey flows through fairly level country. But NE of Iringa, between Mtera and Kidatu, the river cuts through the Southern Highlands in a series of antecedent gorges (Plate 5.17) with numerous rapids. The antecedence was caused by uplift associated with warping and faulting in the Tertiary and Quaternary eras. In the initial stages of uplift extensive deposition took place along the Ruaha Valley upstream of Mtera (Fig. 5.44). This deposition has left thick alluvium

Plate 5.17 The antecedent gorge of the Great Ruaha river, NE of Iringa, Tanzania

deposits for 100 km between Ikorongo and Mtera (1:250,000 Tanzania sheet SB-37-13).

*The Niger River antecedent above Jebba, Nigeria*
Between Yelwa and Jebba the Niger crosses the Kontagora anticline, but since the river was able to maintain its course during the anticlinal upwarp it is antecedent. Above and below the antecedent course the river flows in a broad flood plain over Cretaceous sedimentary rocks. But between Yelwa and Jebba the cover of sediments has been eroded from the axis of the anticline, and the river has cut into the crystalline basement rocks causing a series of rapids and gorges, the upper part of which are now covered by the Kainji Lake.

In Algeria, several rivers follow antecedent cour-

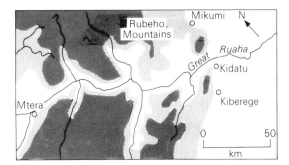

☐ Land over 750 m

■ Land over 1000 m

Fig. 5.44 The Great Ruaha River crossing the Southern Highlands, NE of Iringa, Tanzania

ses through the Atlas Mountains in the form of deep gorges, notably the Rhumel at Constantine and the Kerrata north of Setif.

# Waterfalls and rapids

A waterfall is a sharp break in the channel bed over which the river falls. At the base of the fall the bed may be overdeepened by the hydraulic force of the falling water, swirling round rocks to form a plunge pool. This same process gradually undercuts the face of the fall and causes it to retreat upstream, leaving a gorge downstream from the falls.

Waterfalls originate from a number of causes, of which the main ones are listed below. Some falls are due to only one of these causes, but many have a composite origin. Falls are most common in the upper course, but they may occur along any section of a river. In Africa several major falls are found in the lower course due largely to the complex evolution of the drainage systems (see page 116), and the limited erosive power of the rivers (page 81).

The following are the main causes of waterfalls.

1 An outcrop of hard rock overlying softer rocks in the river bed. This is the most common cause, especially where the resistant outcrop is horizontal or dips upstream. The weaker underlying rock is more easily eroded which causes a steeper gradient and eventually a waterfall, e.g. Kinkon Falls, Guinea (Plate 3.16).

2 Faulting across the river bed. A faultline in the river with weak rocks downstream faulted against resistant rocks will lead to erosion of the weak rocks and the formation of a fall at the faultline. Faulting may also cause a fall directly if a river drops over the edge of a fault scarp, e.g. Kalambo Falls, Zambia (Plate 2.7).

3 Where the river enters the sea at a cliff line. A fall may develop near the mouth of a river if wave erosion cuts back the cliff face or where sea level has fallen, e.g. Lobé Falls, Cameroon (Plate 5.18).

4 Where a tributary hanging valley enters a glacially overdeepened major valley (see page 178).

5 A lava barrier or landslide across a river may initially create a lake (pages 235–36), and a waterfall is likely to form at the overspill from the lake where the river drops over the edge of the barrier, e.g. Lily Falls, Madagascar (Plate 5.19 and Fig. 3.10).

Plate 5.18 Lobé Falls, 7 km south of Kribi, Cameroon, where the Lobé river plunges over rapids directly into the sea

Plate 5.19 Lily Falls, west of Tananarive, Madagascar. Here the Lily falls over the edge of a lava barrier. In the centre, on the skyline, can be seen the cumulo dome volcano of Angavo

6 Where a river falls over a plateau edge, especially when flowing from a high level plain to a lower one, e.g. Augrabies Falls, South Africa (Plate 7.4).

7 Where rejuvenation of a river valley has formed a sharp knickpoint (see page 100).

Rapids are a section of rough, fast-flowing water in a river channel where the bed is steep and rocky. In contrast to a waterfall there is no definite break of slope, but rapids may extend for several kilometres along a river valley. They originate in a number of ways, for example where the rock bar is insufficiently resistant to form a waterfall, or where the resistant rock dips downstream, or where a waterfall has been steadily eroded till it no longer causes a vertical fall. Rapids are common features along many African rivers (pages 81 and 116).

### Victoria Falls (Mosi-oa-Tunya)

Victoria Falls are among the most magnificent of scenic wonders. Here the Zambesi drops 110 m into a deep chasm producing the world's longest sheet of falling water, over $1\frac{1}{2}$ km across (1:5000 Zambia sheets LL 7616 & 7618). As it falls the river sends up a great spray of water visible more than 40 km distant and appearing like rising smoke, hence the name Mosi-oa-Tunya (The Smoke that Thunders).

The chasm into which the Zambesi drops is a narrow cleft at right angles to the course of the river. It forms the first of a series of zig-zag gorges (Fig. 5.45), in total about 100 km long, known collectively as the Batoka Gorge (Plate 2.9). The falls have a composite origin. They are a rejuvenation knickpoint slowly retreating upstream that has been held at an extensive zone of hard basaltic lavas. The form of the falls and the alignment of the gorges is related to erosion along fault lines and joint systems. The zig-zag pattern that has developed is due to the main joint systems in the basalt being diagonally intersected by NE–SW trending faults. The evidence of faulting is especially clear in the second and fifth gorges below the falls.

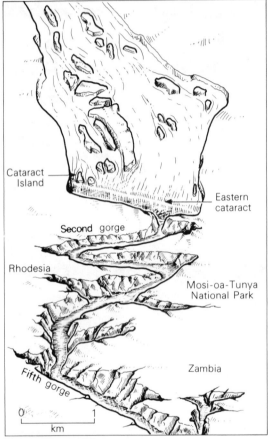

Fig. 5.45  Victoria Falls

The falls have slowly retreated upstream leaving a gorge behind at every stage, and the upstream side of each of these gorges was the former site of the falls. It has been estimated from Stone Age artefacts found near the gorges that the falls were sited at the fourth gorge about 10,000 years ago.

The falls appear to cut back by eroding from one end to another along a line of weakness behind the falls of the time. Modern evidence of this process is visible in the cutting of a trench behind Cataract Island near the Rhodesia bank and along a line parallel to the fault zone of the present second gorge.

Other examples are: Kabarega Falls (Murchison Falls), Uganda, where the upper Nile, some 35 km before entering Lake Mobutu, drops 40 m and cuts a narrow gap less than 6 m wide across basement metamorphic rocks (Fig. 2.11) and Boti Falls, Ghana, where the upper Pawmpawm River drops nearly 20 m over the south edge of the Kwahu Plateau before flowing into the Volta Lake (Plate 1.1: note the picture was taken in the dry season).

*The rapids of the Lower Congo*

Between Kinshasa and the sea the Congo has carved a deep gorge across the folded basement rocks of the Crystal Mountains and produced one of the longest sections of falls and rapids in the world. In this lower course, before entering its estuary at Boma, the Congo plunges over more than 60 rapids and falls, dropping 275 m in about 400 km (Figs. 5.46 and 5.47). Before the Quarternary era the Congo entered the Atlantic near to the present area of south Cameroon. However, uplift along the margins of the continent sealed off its exit and impounded a huge lake in the Congo Basin. It is less

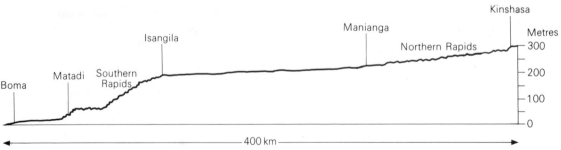

Fig. 5.46  Long profile of the Lower Congo

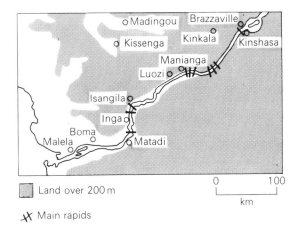

Land over 200 m

✕ Main rapids

Fig. 5.47 The Lower Congo

than 15,000 years since the lake was drained and the new outlet to the sea formed by cutting the present overflow gorge through the Crystal Mountains.

Other examples of rapids: Nile Cataracts, Egypt and Sudan; Felou Rapids, near Kayes on Senegal River, Mali; Bidaga and Popoli Rapids below Soubre, Ivory Coast, on Sassandra River.

## Gorges

A gorge is a deep, narrow river valley with precipitous sides. Gorges are most common in the upper part of river valleys, but under special circumstances they may develop along any section of a river course. In general they occur where the river crosses outcrops of resistant rock, but the actual origin of any one gorge may be due to a number of causes, most of which have already been described:

1. a waterfall retreating upstream, e.g. Batoka Gorge, below Victoria Falls (Plate 2.9);
2. a superimposed gorge where a river is superimposed onto a zone of hard rocks from a former covering rock layer, e.g. Sabaloka Gorge, Sudan (Plate 5.16);
3. an antecedent gorge, where a river has cut across a zone of rocks that is being slowly uplifted, e.g. Great Ruaha, Tanzania (Plate 5.17);
4. a river flowing across (a) an arid area, or (b) a region of limestone rocks. In both cases vertical

Plate 5.20 The Fish River Canyon, Namibia. The structural benches on the canyon walls are clearly visible

erosion may be very great relative to weathering on the valley sides, e.g. (a) Fish River Canyon, Namibia (Plate 5.20) and (b) Manambolo Gorge, Madagascar (Plate 6.10);

5 an overflow gorge cut by a river draining a lake, e.g. Lower Congo Gorge below Kinshasa (page 113);

6 a rejuvenation gorge where a river incises itself into the land surface due to a fall in base level, e.g. Lupata Gorge, Mozambique (Fig. 5.49).

Many gorges have composite origins, for example the Rhumel Gorge at Constantine, Algeria, which is both antecedent and cut across limestone rocks.

Where a river valley is cut in horizontal strata of alternately hard and soft rocks, differential erosion may lead to the development of structural rock benches (Fig. 5.48). Features of this type are most spectacular in steep-sided valleys and gorges formed in arid areas where there is no vegetation to mask their outline. A notable example is the Fish River Canyon in Namibia (Plate 5.20 and Fig. 8.2).

The Fish River flows south to join the Orange from the Khomas Highlands near Windhoek. In the lower part of its course, west of Karasburg, the river has cut a gigantic canyon into the horizontal quartzitic sandstones and shales of the Nama system to reach the underlying Precambrian schist and granite gneiss. In places the canyon is over 500 m deep, while its sides are stepped by a series of huge structural benches (1:500,000 South Africa sheet SE 29/15).

# River valley profiles

The enormous length of rivers results in them flowing across a wide variety of rock type and structure, and this, combined with their great age, produces considerable variation in the long and cross profiles of river valleys from one section to another.

The long and cross profiles of a river valley vary according to whether the river is in the upper course (page 82) or the lower course (page 83), whether it has been rejuvenated (page 96) and whether it is flowing over weak or resistant rocks. In its upper course the cross profile is generally V-shaped, and near its mouth a wide, flat plain. However, if the river has been rejuvenated or superimposed onto resistant rocks near its mouth, the cross profile may be a narrow steep-sided gorge in the lower course.

The cross profile of a river flowing over alternately weak and resistant rocks may alternate between being wide and open on the one hand, and steep and narrow on the other. This is the form of the Zambesi, below Tete, in Mozambique (Fig. 5.49). For 60 km below Tete the river flows in a fairly wide valley bordered by alluvial flats and low hills. In this reach the main rocks are Karoo sandstones. But near Sunga it enters the Lupata Gorge, where it has incised itself, due to rejuvenation, into a resistant outcrop of rhyolites and basalts. The rhyolites show marked columnar jointing. Within the gorge, which is about 15 km long, the Zambesi has a narrow channel and flows between steep vertical cliffs, which towards Bandar rise 300 m above the river. Below Bandar, however, the river crosses soft Cretaceous sediments and

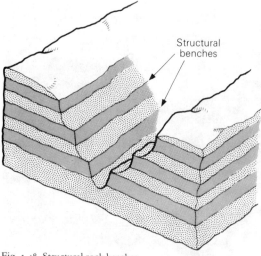

Fig. 5.48 Structural rock benches

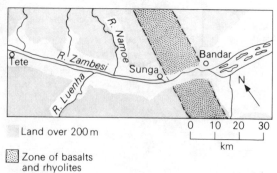

Land over 200 m

Zone of basalts and rhyolites

Fig. 5.49 The Lupata Gorge, Mozambique

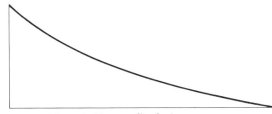

Fig. 5.50 Theoretical long profile of a river

takes on a broad braided channel in a wide, open valley.

The long profile of a river tends towards being concave upwards (Fig. 5.50), but few rivers have a profile as regular as the theoretical curve of Fig. 5.50. In fact, due to variations in rock resistance and changes in base level long profiles have many sections that are convex.

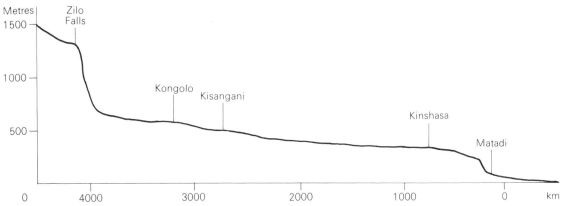

Fig. 5.51 Long profile of the Congo

Plate 5.21 The wide channel of the Congo river at Kisangani, Zaire

In contrast to rivers in temperate latitudes, tropical rivers, with the fine nature of their loads (page 81), are far less successful in eroding hard rock outcrops along their beds, with the result that rapids are a common feature. The inability to wear away rapids accounts for many of the irregularities in African river profiles. The numerous falls and rapids in the lower reaches of rivers reflect the recent negative movements in base level (page 12) during the Pleistocene period. Because of this, long profiles tend to show a marked convexity in their lower course (Fig. 5.51). Major falls are often giant knickpoints (page 100) slowly retreating upstream, for example Victoria Falls and Augrabies Falls. They indicate the point where the river flows from an older high level plain to a younger low level plain (Chapter 7).

Between the various falls and rapids rivers in Africa generally flow for long distances over wide flat plains at very gentle gradients, especially where they cross areas of large-scale deposition, such as the Congo Basin. In such reaches the rivers develop broad shallow channels and features characteristic of the lower course, as in the Barotse Plain of the Zambesi upstream from Victoria Falls (Plate 5.6).

These factors help to explain the convex nature of the long profile of many African rivers, including very large ones like the Congo. This huge river, which with its tributaries drains an area of 3.7 million square kilometres, has an average gradient of less than 7 cm per km between Kisangani (Plate 5.21) and Kinshasa. But below Kinshasa, in the rapids section (Fig. 5.46), the gradient increases abruptly to nearly 70 cm per km.

# References

BERRY L. and WHITEMAN A. J., 'The Nile in the Sudan', *Geographical Journal*, Vol. 134, pp. 1–37, 1968.

HILTON T. E., 'River Captures in Ghana', *Bulletin of the Ghana Geographical Association,* Vol. 9, No. 1, pp. 13–24, 1964.

HUNTER J. M., 'Aspects of the Erosional History of the Upper Birim Basin, Ghana', *Journal of the West African Science Association,* Vol. 5, pp. 108–125, 1959.

JATZOLD R., 'The Kilombero Valley', Ifo-Institut für Wirtschaftsfurshung, München, *Afrika-Studien*, Nr 28, 1968.

NEDECO (Netherlands Engineering Consultants, The Hague), *River Studies and Recommendations on Improvement of Niger and Benue*, North Holland Publishing Company, Amsterdam, 1959.

PARTRIDGE T. C. and BRINK A. B. A., 'Gravels and Terraces of the Lower Vaal River Basin', *South African Geographical Journal*, Vol. 49, pp. 21–38, 1967.

PUGH J. C., 'River Captures in Nigeria', *Nigerian Geographical Journal*, Vol. 4, No. 2, pp. 41–48, 1961.

ROBERT M., *Le Congo Physique* (3rd Edition), Chap. 4: Évolution du Relief et du Système Hydrographique, 1946.

TALJAARD M. S., 'Morphological and Hydrological Aspects of the Tulbagh–Swellendam Mountain Foreland', *South African Geographical Journal*, Vol. 33, pp. 38–52, 1951.

TEMPLE P. H. and SUNDBORG A., 'The Rufiji River, Tanzania: hydrology and sediment transport', *Geografiska Annaler*, Vol. 54A, pp. 345–368, 1972.

TRICART J., *The Landforms of the Humid Tropics, Forests and Savannas*, Chapter 2, Longman, 1972.

# CHAPTER SIX

# The influence of rocks on landscape

## The relationship between rocks and landscape

In many parts of Africa there is often a close relationship between rocks and landscape. Some rocks, such as quartzites, are more resistant to erosion than others and therefore form uplands, while limestones, because of their chemical characteristics, sometimes give rise to a special type of scenery called 'karst' (page 123). But this is not to suggest that each rock forms its own particular type of landscape, for the influence that individual rocks may have on relief depends on several important factors, none of which should be considered separately since all are closely interrelated.

1　Rock hardness: hard rocks are more resistant to erosion than soft rocks, hence rocks like quartzite, granite and gabbro often form upland areas, while soft rocks like clay and shales form lowlands. Rock hardness may also influence the steepness of slopes. Soft rocks tend to produce gentle slopes, while hard rocks often form steep slopes.

2　Rock permeability: permeable rocks like sandstone and limestone allow surface water to percolate through them, so reducing runoff and stream erosion. For this reason permeable rocks may form uplands in contrast to clays which beside being soft are also impermeable and thus susceptible to stream erosion (see page 7).

3　Rock jointing: joints (page 6) provide access for the agents of weathering and erosion, and hard rocks which are well-jointed will eventually be worn down in this way. However, rocks with well developed joints are also permeable, which will tend to reduce surface erosion. Limestones fall in this group. Rock jointing is also a major factor in determining the shape and size of certain individual landforms, notably inselbergs (pages 140–44) and karst features (pages 125–28).

4　The nature of the adjacent rocks: the extent to which a particular rock will form upstanding relief will depend not only on its resistance, but also on the relative resistance of the surrounding rocks. For this reason the same rock may appear more resistant in one place than in another (page 54).

5　Structure: where alternately hard and soft beds have been tilted the effect of differential erosion will be to wear away the weaker rocks while leaving the more resistant rocks as uplands. The nature of the landscape will then depend on how

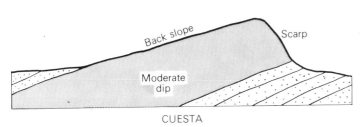

Fig. 6.1 A cuesta and a hogback

steeply the rocks dip (page 2). Moderately dipping strata tend to form a cuesta scarpland topography. A cuesta is a ridge or upland with a steep scarp slope and a gentle back slope. When the dip is very steep the result is a ridge with steep slopes on both sides, called a hogback (Fig. 6.1). Horizontal and very gently dipping strata produce a plateau-like relief, especially where it is capped by a resistant rock layer. If the resistant rock is broken through by erosion, smaller plateau areas known as mesas develop. The dissection of a mesa leads to the formation of isolated flat-topped hills called buttes (Plate 3.15). In the tropics buttes and mesas may also be capped by a hard laterite duricrust (page 68). As the resistant cap is gradually eroded away the hill may develop a cone-shaped outline.

6 Climate: the same rock may in one region form an upland and in another a lowland, due to differences in climate. In hot deserts the dry climate makes limestones relatively resistant, but in equatorial regions the same rock is soon worn down by solution (page 127).

7 The stage of erosion: in areas where the geomorphic cycle (page 12) has been able to run its course without interruption, so that a late stage of erosion is reached and a level plain is formed, relief may bear little relation to differences in rock resistance (page 137).

### Examples

*Bandiagara Plateau, Mali*
South of the inland delta of the Niger the land rises gently to form the great Bandiagara Plateau (Fig. 6.2). The plateau, with its high east-facing scarp (Plate 6.1) rising 250 m from the flat sand-covered

Plate 6.1 The almost vertical escarpment of the Bandiagara Plateau in northern Mali. Notice the steep talus slope rising behind the village

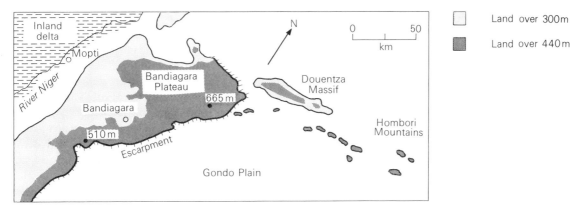

Fig. 6.2  The Bandiagara Plateau

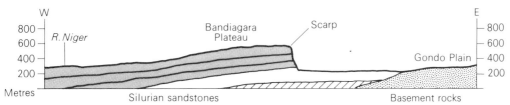

Fig. 6.3  Cross-section through the Bandiagara Plateau

Gondo Plain, is composed of hard, gently dipping Silurian sandstones (Fig. 6.3). It is part of the south-eastern edge of the vast Taoudene structural basin (Fig. 1.14) that underlies much of Mali and Mauritania.

The Bandiagara scarp is a good example of an erosion scarp. Others are the Gambaga and Kwahu scarps in Ghana. An erosion scarp is caused by the gradual wearing back of a hillside or plateau edge. A very gentle slope may thus be transformed into a steep scarp. An erosion scarp should not be con-fused with a fault scarp (page 15). The Bandiagara sandstone is especially resistant because of the impregnation of silica, and this, together with the

Plate 6.2  Some of the mesas and giant buttes that make up the Hombori Mountains, Mali

limited weathering of the semi-arid climate, is largely responsible for the very steep nature of the scarp. The chief zones of weakness in the sandstone are the joint systems with their dominant south-western alignment. Several landforms on the plateau show the influence of joint control. Streams, for example, tend to follow a south-westerly course rather than flow directly down the backslope to the Niger.

The Bandiagara plateau is continued north-east by the huge mesa of the Douentza Massif and a series of gigantic buttes that culminate in the Hombori Mountains rising more than 600 m above the plain (Plate 6.2). (1:200,000 Mali IGN sheet ND-30-XV.)

Other examples of plateaus with steep scarps and residual buttes are: Gilf Kebir Plateau, SW Egypt (page 165); Adrar, north Mauritania, and Assaba and Tagant Plateaus, south Mauritania, north of Sélibabi.

### The granite massif of Andringitra, Madagascar
Andringitra is a vast north–south trending massif of Palaeozoic granites and syenites that rises 2658 m in Mount Boby, second highest point in Madagascar. It is composed of one large upland mass with some smaller extensions to the north. The massif rises nearly 800 m above the adjacent quartzites and shales into which it was intruded. In places, especially the south-east, the bounding slopes are almost vertical. The summit area of the main upland is a region of huge rounded boulders and deep clefts, above which rise six granite peaks. These peaks, and the smaller uplands near Ambalavao (Plate 6.3), show marked exfoliation (page 65) on the smooth slopes of the granite. Chemical weathering attacks the massive rock more or less equally all over and this gives rise to a generally rounded topography.

### The quartzite uplands of East Transvaal, South Africa
South of Barberton, on the Swaziland border, differential erosion of the heavily folded Lower Precambrian rocks has resulted in a series of uplands and intervening valleys (1:50,000 Swaziland/South Africa sheets 2531 CC, 2531 CD). The uplands are

Plate 6.3 Granite uplands near Ambalavao, Madagascar. This is one of several outliers that extend north from the main Andringitra Massif

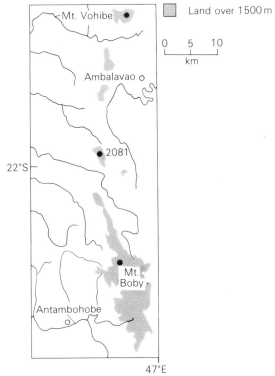

Fig. 6.4 Andringitra Massif, SE Madagascar

formed in resistant quartzites of the Moodies series, while the rivers, especially the Lomati and its tributaries, have carved longitudinal valleys in the softer shales, producing a trellis pattern (Fig. 6.5). Both Saddleback and Makhonjwa ridges are former synclines. The relief has been inverted and the original downfolds are now major uplands (Fig. 6.6). Intruded into these ancient rocks are dykes of various ages, many of which, due to limited resistance, have been deeply weathered. Streams that cut transversely across the ridges often follow the line of weathered dykes. This is especially apparent in the gorges cut by the Mhlume and Ugutugulo Rivers, where they cross the Makhonjwa Range on either side of Ufafa Peak (Plate 6.4).

Other examples of quartzite uplands are the Nimba Mountains, on the Liberia–Guinea border, with the high peak of Mont Richard Molard (1752 m) (differential erosion is evident in a number of river valleys carved along outcrops of weaker schists, notably the upper Ya at about 700 m); Serra da Chela, SW Angola (Fig. 5.31); and Itremo Massif, central Madagascar.

Plate 6.4 Vertical air view of part of the quartzite Makhonjwa Range on the Swaziland border, SE of Barberton, South Africa. The Mhlume and Ugutugulo rivers have eroded their valleys along the line of weathered dolerite dykes

The area of this photo is that bounded by the dotted line in Fig. 6.5

121

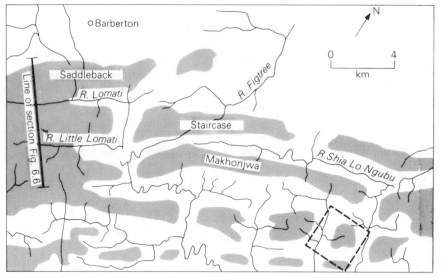

Fig. 6.5 The quartzite uplands south of Barberton, Transvaal, South Africa

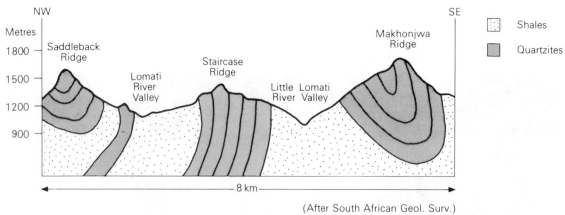

(After South African Geol. Surv.)

Fig. 6.6 Cross section of the quartzite uplands

### The scarplands of SE Nigeria

In SE Nigeria differential erosion of gently dipping sandstones, clays and shales has given rise to a scarpland relief (1:50,000 Nigeria sheets 301 NE & 301 SE). The Tertiary sediments east of the Niger have, in the past, been folded into an anticline, of which only the western limb now remains (Fig. 6.7). The hard sandstones of this western limb form the Udi Cuesta, an upland with an east-facing scarp that separates the Cross River Plains from the Anambra Plains. In places, the scarp rises to nearly 450 m, or

150 m above the adjacent lowlands (Plate 6.5). In the south, to the west of the upper Imo and Mamu Rivers, Eocene sandstones form a second scarp on the east side of the Awka Uplands (Fig. 6.8).

Dissection has cut into both escarpments, and many areas have been severely gullied (page 75). In several places spurs and outliers project into the lowlands in front of the scarp. An example of one of these outliers is Juju Hill, south of Enugu, which is protected by a capping of laterite duricrust.

Plate 6.5 The scarp of the Udi Cuesta rising behind the Uwani residential district of Enugu, Nigeria

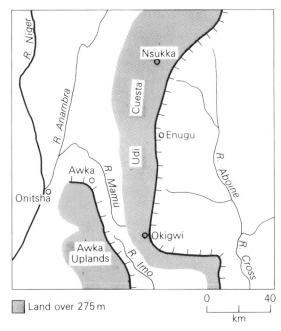

Land over 275 m

0      40
km

Fig. 6.7 Udi Cuesta, Nigeria

*Guelb Er Richat, Mauritania*

Guelb Er Richat is a vast circular depression in the Adrar region of Mauritania. It is about 50 km in diameter and lies some 550 km east of Nouadibou. It has been formed by the differential erosion of alternate outcrops of updomed Palaeozoic schists and quartzites (1 : 80,000 Mauritania IGN Er Richat special sheet). The quartzites are considerably more resistant than the schists, and dissection has left them as concentric ridges surrounding a central core, notably Rich Amou er Rejel and Rich Seriz (Fig. 6.9). The weaker schists, however, have been eroded and now form the annular depressions occupied by dry river channels and salt flats.

## Karst scenery.

The influence of rock type on landforms is nowhere so clearly visible as in those areas where limestones and dolomites outcrop. These rocks, because of

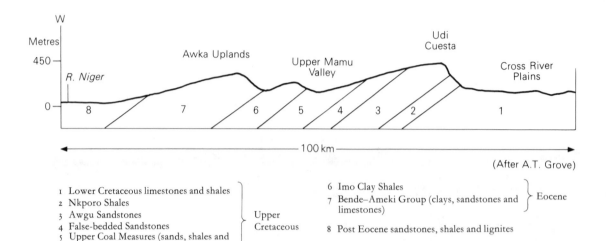

Metres
450 —

R. Niger

Awka Uplands

Upper Mamu
Valley

Udi
Cuesta

Cross River
Plains

0 —

8    7    6    5    4    3    2    1

⟵——————— 100 km ———————⟶

(After A.T. Grove)

1 Lower Cretaceous limestones and shales
2 Nkporo Shales
3 Awgu Sandstones
4 False-bedded Sandstones
5 Upper Coal Measures (sands, shales and coal)
⎫ Upper Cretaceous

6 Imo Clay Shales
7 Bende–Ameki Group (clays, sandstones and limestones)
⎫ Eocene

8 Post Eocene sandstones, shales and lignites

Fig. 6.8 Cross section of the Udi Cuesta and Awka Uplands

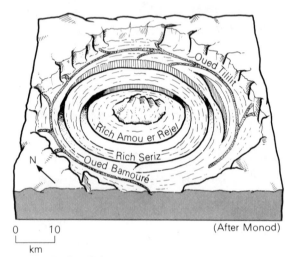

Oued Tililt

Rich Amou er Réjel

Rich Seriz

Oued Bamouré

N

0    10
⎣____⎦
km

(After Monod)

Fig. 6.9 Guelb er Richat

their solubility in water containing carbon dioxide, sometimes form a very distinctive type of landscape known as 'karst'. The name karst is derived from the limestone plateau area of Yugoslavia, where a large number of very individual landforms have developed due to solution. The formation of similar features in other areas of calcareous rocks has led to the term karst being applied to a number of regions throughout the world in which some of the landforms characteristic of the Yugoslavian limestone can be identified.

Limestones are composed mainly of calcium carbonate, $CaCO_3$, and dolomites of calcium mag-nesium carbonate, $CaMg(CO_3)_2$. Both rocks can be dissolved in carbonic acid, that is, water charged with carbon dioxide, $CO_2$, and thus converted into calcium bicarbonate, $Ca(HCO_3)_2$, which is soluble and can be carried away by groundwater in solution. This process is known as carbonation solution.

Some of the carbon dioxide is absorbed from the air by rainwater but the greatest concentration occurs in the soil and vegetation cover, and is taken up by ground water as it percolates downwards. Also important in the solution process is the occurrence of organic acids from decaying vegetation.

But not all calcareous rocks give rise to karst scenery, since other factors are important as well as the solubility of the rock, especially the following.

1 The rock should be hard and well-jointed to make it permeable. It is along the joints that acidic ground water is able to percolate and concentrate its action. By this means joints are enlarged and crevices formed. Solution is parti-cularly active where joints are close together or intersect each other.

2 Climate. Moderate to abundant rainfall is needed for the full development of karst landforms. Where such landforms occur in areas of low rainfall they are probably due to a more humid past climate. Karst is found in regions ranging from equatorial to temperate. In the tropics high temperatures and rainfall greatly increase the rate

of solution, which results in a special tropical karst.

3 The water table should be far enough below the surface so that water is always able to move down through the rock.

Where one or more of these factors is absent it is unusual to find any extensive development of karst scenery. For this reason there are surprisingly few areas of karst in relation to the widespread outcrops of limestone and dolomite.

### Chalk

Chalk is a very pure, fine-grained type of limestone composed almost entirely of calcium carbonate. It outcrops in a number of places in northern Africa, for example in Egypt to the west of the Nile at Aswan, but nowhere in the continent does it give rise to the 'chalk scenery' typical of southern England and northern France.

### Major characteristics of karst scenery

1 Surface drainage is intermittent or absent: streams that do occur rarely flow for long distances, but disappear underground through solution hollows.

2 Outcrops of bare, rugged rock, and steep-sided dry valleys.

Plate 6.6 Rough limestone pavement north of Ifrane, Middle Atlas, Morocco. Several large areas of pavement, some more than $\frac{1}{2}$ km across, occur south of Dayete Ifrah, 20 km NE of Ifrane

3 Numerous solution depressions and residual hills of various sizes.

4 A subterranean network of caverns and watercourses.

The action of carbonation solution working along bedding planes and joints in the rock leads to the formation of a series of landforms of which the most important are described on pages 125–28. However, due to differences in lithology, structure and climate no one area of the world includes all these various landforms. An important point to note when studying karst scenery is the variety of names applied to many landforms due to local terminology in different countries. Thus, the corresponding French term for karst is causses, and for obvious reasons this name is used in some African countries.

## Landforms in karst areas

### Limestone pavement

This is a bare rock surface, criss-crossed by numerous grooves where solution has worked along the joints (Plate 6.6). It appears as a number of narrow ridges (lapiaz, clints) divided by clefts (lapiés, grikes), which may be up to a metre deep. Some areas of pavement cover several hundred square metres.

### Doline

This is a shallow depression or hollow with gently sloping sides and generally circular or oval in plan. Dolines vary from a few metres to several hundred metres in diameter and usually occur in large groups. They originate from water percolating underground at the intersection of major joints.

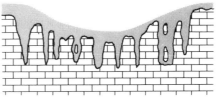

Fig. 6.10 Doline

The rock is slowly dissolved and a small basin formed, which is gradually enlarged (Fig. 6.10).

## Collapse doline

This contrasts with a solution doline in having steep sides. It is caused by the collapse of part of an underground cave where the roof is particularly thin. This type of doline sometimes occurs in chains above the course of a subterranean stream (Fig. 6.11).

Fig. 6.11 Collapse doline

## Uvala

This is a large closed depression, intermediate in size between a doline and a polje, with a diameter generally greater than 500 m. It is formed by the coalescence of several dolines. The floor is often hummocky, consisting of several interconnected hollows.

## Polje

This is a very large, shallow, steep-sided depression with a generally flat floor. It may be many kilometres in size. The flat floor is often emphasised by the deposition of terra rossa, a red clay material, which may form an impermeable layer and lead to flooding after heavy rains. Small residual hills, known as hums, form significant features in some poljes, and depending on the depth of the water table a polje may hold a temporary or permanent lake. Many of these giant depressions coincide with structural basins formed by folding or faulting. The

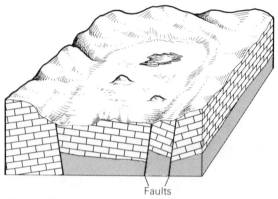

Faults

Fig. 6.12 Polje

large-scale solution involved in their origin has thus been guided by fault lines or fold axes (Fig. 6.12).

## Sink or ponor (plural: ponore)

This is a deep hole with nearly vertical sides leading to an underground cave system. Some result mainly from surface solution along joints and some from a combination of solution and subsurface collapse. Many are located on the floors of dolines, uvalas and poljes. A sink where a river disappears underground and is lost to surface drainage is sometimes called a swallow hole.

## Dry valley

This is a valley with no permanent stream and often with steep almost vertical sides. Most are caused by a gradual lowering of the water table, which itself is often due to the entrenchment of major streams into the limestone. As the water table falls small streams become intermittent first and finally vanish. Some dry valleys may, however, result from roof collapse along a subterranean watercourse. Steep valley sides are caused by the vertical joint planes in the rock.

## Blind valley

This is a valley closed at the lower end by a rock wall, at the base of which is a sink where a stream vanishes underground. Such a valley is usually the result of a lowering of the river bed by solution around the sink, and subsequent retreat upstream. Some may be due to partial underground collapse. Many blind valleys are also dry valleys (Fig. 6.13).

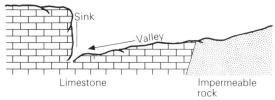

Fig. 6.13 Blind valley

### Limestone gorge

This is a deep, steep-sided valley formed by a large river, with its headwaters beyond the limestone area, steadily entrenching itself into the landsurface. The steep cross profile is partly a result of the nature of the rock jointing. Also important is the reduced run-off, due to infiltration, which limits the extent to which mass wasting and downwash can wear back the valley sides.

### Tropical karst residuals

High annual rainfall combined with large quantities of organic acid from decaying vegetation increase the rate of solution and give rise to a distinctive tropical karst, of which there are two main varieties.

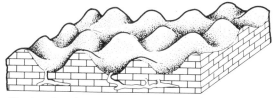

Fig. 6.14 Cone karst

1 Conekarst (kegelkarst): a series of cone-shaped hills separated from each other by multi-sided closed depressions, producing a hummocky landscape. (In Jamaica this scenery is known as 'cockpit country'.) The hollows between the hills, which may be over 100 m deep and up to 400 m across, seem to coincide with those parts of the rock that are strongly jointed and more susceptible to solution. The depressions are deepened by increased solution caused by concentrated rainfall and a richer vegetation growth (Fig. 6.14).

2 Towerkarst (turmkarst): a series of steep-sided hills undercut at the base, and separated from each other by areas of level, clay-covered plain. These towers seem to form in hard limestone with marked vertical jointing. The undercutting of the towers and the smooth nature of the plains are the result of regular flooding, which would suggest that the water table is very near the surface (Fig. 6.15).

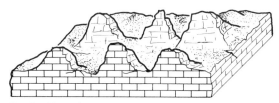

Fig. 6.15 Tower karst

### Limestone cavern

This is a natural subterranean chamber in the rock, usually joined to the land surface by a system of interconnecting shafts and galleries. Caves formed in limestone are larger and more numerous than those formed in other environments. Some limestone caves are dry, some have underground streams flowing through them. The main factors in cavern formation are solution, together with mechanical erosion by underground streams, and also occasional roof collapse. Ground water circulating in the zone below the water table dissolves the limestone and forms cavities by attacking joints and bedding planes. As the water table falls, vadose water (Fig. 1.9) percolating down through the rock, continues the process, enlarging the cavities. Finally, rivers taking a subterranean course may increase the size of the cavities by stream erosion. Hence limestone caves often appear at varying levels in the rock, above and below the water table, while their shape and the pattern of the underground galleries generally reflects the influence of joint control. Streams that flow through caves may eventually emerge at the surface near the base of the limestone, and the point at which they reappear is called a resurgence. In many caves there are stalactites, stalagmites and other calcite deposits. A stalactite is a mass of calcite (crystalline calcium carbonate) hanging from the roof of a cave, while a stalagmite grows up from the floor. These deposits form as water seeps through cracks in the rock and drips from cave ceilings and walls. Carbon dioxide escapes from the water causing some of the dissolved calcium bicarbonate to change back into calcium carbonate.

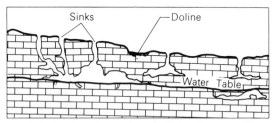

Fig. 6.16 Cave formation at, above and below the water table

The landforms described above are not peculiar to limestones and dolomites, for caves, depressions and lapiés are sometimes found in other rocks such as sandstones, quartzites and basalts. Certain types of sandstone are easily modelled by water, and this can lead to the formation of caves. In South Africa the Cave Sandstone (Fig. 1.6) was named from its caves, such as those in the Witberge, 30 km north of Barkly East, Cape Province. Active chemical weathering is the major cause of the lapiés and depressions formed on many quartzite and basalt outcrops in the equatorial zone, for example the quartzite massif of Ibity, 25 km south of Antsirabe in Madagascar.

### Dolines, poljes and dry valleys in the Middle Atlas, Morocco

In the Middle Atlas Mountains, especially between Ifrane and Khenifra (Fig. 6.17), a karst landscape of numerous solution hollows has been formed in the folded and faulted Jurassic limestones and dolomites (1 : 100,000 Morocco IGN sheet NI-30-VII-2).

Dolines and collapse dolines of various shapes and sizes are widespread, while the influence of faulting has concentrated solution into many down-faulted sectors and led to the formation of poljes.

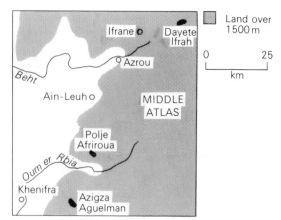

Fig. 6.17 Main karst region in the Middle Atlas Mountains, Morocco

However, few of the limestones are pure enough or thick enough to favour the development of any underground drainage. Also, much of the karst may be fossilised since present rainfall totals are not sufficient to encourage very active solution.

Two typical poljes are Afriroua (Plate 6.7) and Ksouatene west of the upper Ouiouane Valley, about 20 km south of Ain-Leuh (Fig. 6.18). Ksouatene is open at one end and contains three ponore, while at the east end of Afriroua is an aguelman, or shallow lake, which lies at the lowest part of the polje. The floors of the depressions are

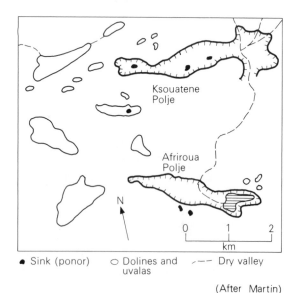

(After Martin)

Fig. 6.18 Poljes, dolines and uvalas in the Middle Atlas

Plate 6.7 Afriroua Polje, Middle Atlas, Morocco. Note the flat grass-covered floor and the steep wooded sides

flat and mostly covered by clay deposits. In the middle of Afriroua are two small hums. The short grass of the polje floors contrasts with the woodlands that cover the surrounding uplands. The sides of both depressions are quite steep and are aligned with the major faults of the area (Fig. 6.19).

A similar but larger polje is Dayete Ifrah, 20 km NE of Ifrane, around which are numerous dry valleys, blind valleys and dolines, such as Regada doline on the SW side.

in the east (Fig. 6.20). This part of Madagascar is semi-arid, rainfall 400 mm and less, with the result that the Mahafaly karst is developing only very slowly. Over the plateau there is no surface drainage except the river Linta in the south, which flows for a short while each year in its gorge-like valley. However, the existence of a subterranean drainage is evident from the presence of springs and resurgences in the coast region. Lake Tsimanampetsotsa is thought to be fed by a number of these

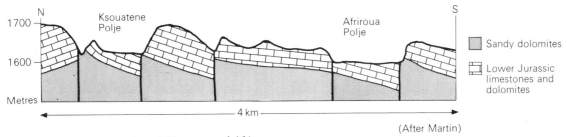

Fig. 6.19 Geological section through Ksouatene and Afriroua poljes

*Dolines and sinks on the Mahafaly Plateau, Madagascar*
The Mahafaly Plateau lies in SW Madagascar, between the rivers Onilahy and Menarandra (1:500,000 Madagascar sheet 11). It is formed of Eocene limestones and is nearly 200 km from north to south, with an average width of 50 km. The height varies from 100 m in the west to over 350 m

resurgences. Also, several caves and sinks have been discovered, for example Andavaka Cave, Mitoho Cave and Lavaboro Sink. Lavaboro Sink (or aven) is 115 m deep and about 20 m wide at the mouth (Fig. 6.21). After a drop of 70 m the vertical shaft is interrupted by a projecting step.

The dominant karst features of the plateau are

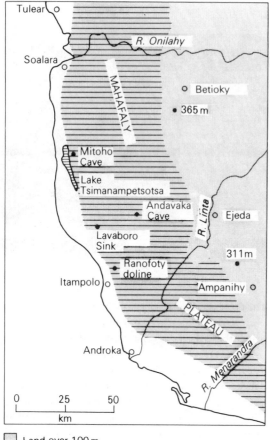

Fig. 6.20 The karst plateau of Mahafaly, SW Madagascar

Land over 100 m

Limestone plateau

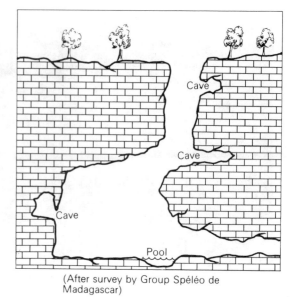

(After survey by Group Spéléo de Madagascar)

Fig. 6.21 Lavaboro Sink, Mahafaly Plateau, Madagascar

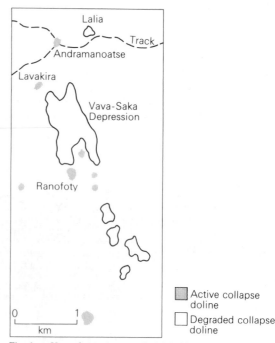

Active collapse doline

Degraded collapse doline

Fig. 6.22 Karst depressions on the Mahafaly Plateau, Madagascar

dolines, uvalas and collapse dolines. The dolines and uvalas are most widespread in a north–south central zone. They vary in diameter from very small to more than 500 m, but few exceed 30 m in depth. The flat floors of the depressions carry little vegetation in contrast to the very thick cover of stunted trees and bushes over the plateau surface.

But it is collapse dolines that are the most remarkable features of Mahafaly. These steep-sided cauldron-like depressions are often known as avens. (Aven is a French term, meaning a shaft open to the sky, but its use is best restricted to sinks, like Lavaboro.) They are especially concentrated in a narrow belt, about 15 km wide, that lies on the west of the plateau and north of Itampolo. More than 50 collapse dolines exist in this area, varying in diameter from 10 m to more than 100 m. Some exceed 50 m in depth, and many have almost vertical walls. One of the largest is Ranofoty, which is about 225 m wide and 70 m deep (Fig. 6.22). The floor of

Plate 6.8 The high vertical west wall of the Ankarana limestone plateau, NW Madagascar

each cauldron is mostly covered by large blocks and there is usually a small pool. This pool, governed by the level of the water table, helps maintain the active development of the collapse doline. Once the water table falls and the pool disappears, the depression becomes infilled and the sides are worn down giving it the appearance of a normal solution doline. Several of these degraded collapse dolines occur adjacent to the active ones. They are apparent from their shallow depth and larger size (Fig. 6.22). The huge depression of Vava-Saka is thought to have resulted from the coalescence of several degraded collapse dolines.

### Towerkarst, caves and underground drainage on the Ankarana Plateau, Madagascar

The Jurassic limestone Ankarana Plateau is smaller than Mahafaly, and lies in the far NW of Madagascar, about 25 km north of Ambilobe (1 : 100,000 Madagascar sheet U 32). It is dissected by several narrow, steep-sided corridors and shows evidence of very rapid solution for a seasonally dry climate. Most of the 200 cm annual rain falls between November and March (Fig. 6.23). On the

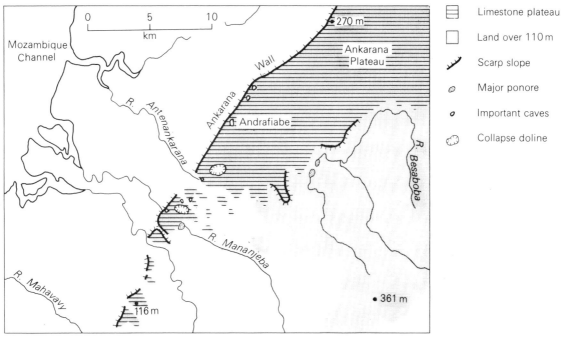
Fig. 6.23 Ankarana Plateau, NW Madagascar

west the plateau is bounded by the 200 m high fault scarp of the Ankarana Wall (Plate 6.8).

The landscape is strongly controlled by faulting and the joint pattern. The main alignment of the joints is NW–SE and this has helped in the dissection of the plateau. The surface of Ankarana has been etched into a vast area of bare rock pinnacles, known in Madagascar as 'tsingy'. In the south, solution guided by fractures and joints has separated the pure hard limestone into smaller blocks and towers that have been fashioned to form a tropical tower karst. Many towers have sharp-edged summits and steep lower sides. They rise to between 50 and 100 m.

The plateau shows no trace of surface drainage and there are relatively few dolines. However, there are a large number of ponore as well as two large shallow depressions, which seem to be related to the extensive subterranean cave network. One of these depressions is 800 m in diameter, and may be a form of ancient collapse doline.

East of the plateau several streams disappear into ponore, and the most remarkable of these is the Mananjeba which follows an underground course for 1½ km. All subterranean water seems to flow SW, and at least one river, the Antenankarana, has its source in a resurgence.

The karst is riddled with caves at different levels, some of which are now dry and some still active with flowing water. Andrafiabe cave system (Plate 6.9) is closely controlled by joints and faults and its layout is parallel to the main fault scarp (Fig. 6.24). Nearly 10 km of galleries, chambers and corridors have been explored, many of them with numerous stalactites and stalagmites.

*Conekarst near Kissenga, Congo*
Dissection along the edges of the Kissenga Plateau in southern Congo (Fig. 5.47) has caused the evolution of a tropical conekarst scenery. The plateau, which is largely composed of dolomitic limestones capped by resistant sandstones of the Mpioka series, lies about 50 km SW of Madingou and rises to about 400 m (1:50,000 Congo IGN sheet SB-33-2-3b).

At the edges of the plateau, where the limestone

Plate 6.9 Entrance to Andrafiabe Cave, Ankarana Plateau

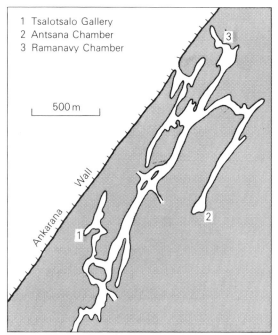

1 Tsalotsalo Gallery
2 Antsana Chamber
3 Ramanavy Chamber

500 m

Ankarana Wall

(After survey 1964–1965 by Group Spéléo de Madagascar)
Fig. 6.24 Plan of Andrafiabe Cave

Karst also exists in west Congo, south of Kibangou, along the Loubetsi Valley, a tributary of the Kouilou. Here there are numerous large shallow dolines, most of them over 500 m in diameter, while near Yama there is a further area of conekarst (karst à pitons).

### The Manambolo Gorge, Madagascar

In the western part of central Madagascar the Manambolo River, flowing west to the Mozambique Channel, has entrenched itself into the limestone massif of the Causse de l'Antsingy in a narrow

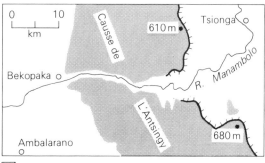

Land over 300 m

Scarp

Fig. 6.26 Manambolo Gorge, Madagascar

outcrops and the overlying sandstone has been removed, there are numerous small conical hillocks interspersed by round dish-like depressions (Fig. 6.25). The steep-sided cones have been grooved and fretted by solution and are mostly less than 50 m high. Although it is a latitude of equatorial rainfall the absence of regular surface drainage on the limestone has produced a grassland vegetation over this 'cockpit' landscape.

gorge several hundred metres deep and over 20 km long (Fig. 6.26). The walls of the gorge rise almost sheer from the water's edge (Plate 6.10). The headwaters of the river lie to the east, away from the limestone outcrop and in an area of higher rainfall.

Fig. 6.25 View of part of the area of small dolomite cones on the SW edge of the Kissenga Plateau, Congo

Plate 6.10 The limestone gorge of the Manambolo River, central Madagascar

Plate 6.11 Stalactites and stalagmites in Matupi Cave, Mt Hoyo, Zaire

Much of the underground drainage of the Causse eventually finds its way into the Manambolo Gorge, having originally disappeared into numerous ponore and dolines (1:100,000 Madagascar sheet G 47). Along the walls of the gorge are several caves, for example Anjohikinakina.

*Other examples of limestone scenery*

Large areas of limestone pavement with numerous gorges, sinks, dolines and other depressions lie some 10 km inland of Tanga in Tanzania. The karst is deeply dissected and is especially widespread near Kiomoni and Kange, either side of the Mkulumuzi valley (Tanzania 1:50,000 sheet 130E/1).

There are many dolines and dry valleys on the dolomite Kaap Plateau of northern Cape Province, South Africa (Fig. 8.2).

Several sinks and caves occur in the Lomagundi dolomite of the Hunyani Hills 135 km NW of Salisbury, Rhodesia. One of the most famous sinks is the Wonderhole which, with its neighbours, lies 6 km west of Sinoia.

*Limestone caves in South Africa and Zaire*

Although caves form a significant feature of karst development they do not by themselves represent karst scenery, as is apparent from Cango Caves, South Africa, and Hoyo Caves, Zaire, both of which occur in areas otherwise devoid of limestone scenery.

Cango Caves lie about 27 km north of Oudtshoorn (Fig. 2.22), at the foot of the Swartberg in the Cape Ranges. The total length of cave passages and galleries is over 3 km, and there are 80 caverns in all, most of them covered with innumerable stalactites and other formations. The caves were formed in a zone of limestone that in the past was shattered by faulting. Considerable solution pockets on the cave walls, together with a lack of any evidence of stream erosion, seem to suggest the caves were formed below the water table.

Mount Hoyo is a fault block upland in NE Zaire, about 40 km SW of Irumu (Fig. 2.8). Along its 300 m high west scarp there are more than 20 caves. The caves, of which Matupi (Plate 6.11) is the most significant, occur at several different levels in the horizontal Precambrian limestones. They were almost entirely formed by ground water action below the water table, and then later raised up by faulting. The rectangular pattern of chambers and passages within the caves shows strong evidence of joint control. Some of the caves, such as Manzenzele and Bertha, contain beautiful collections of stalactites and stalagmites.

Other examples of limestone caves are: Sof Omar Caves, near Goba in SE Ethiopia, where the river Web, an upper tributary of the Juba, follows an underground course for about $1\frac{1}{2}$ km; Salanga Caves, a series of stalactite and stalagmite chambers extending underground for 2000 m, in Kasai Province, Zaire, some 20 km NE of Gandajika; and Tikjda Caves, a group of five caverns (notably Grotte de Glace) in the Djurdjura Mountains, Algeria, 25 km NE of Bourira.

## References

COOKE H. J., 'A Tropical Karst in North East Tanzania', *Zeitschrift für Geomorphologie*, Vol. 17, pp. 443–459, 1973.

DAVEAU S., 'Recherches Morphologiques sur la Région de Bandiagara', *Mémoires de l'Institut Français d' Afrique Noire*, No. 56, 1959.

DE SAINT-OURS J., 'Les Phénomènes Karstiques à Madagascar', *Bulletin de Madagascar*, No. 160, pp. 743–763, 1959.

ILOEJE N. P., 'The Structure and Relief of the Nsukka-Okigwi Cuesta', *Nigerian Geographical Journal*, Vol. 4, No. 1, pp. 21–40, 1961.

KING L. C., 'The Geology of the Cango Caves', *Transactions of the Royal Society of South Africa*, Vol. 33, pp. 457–468, 1952.

MARTIN J., 'Les Poljés du Causse d'El Hammam, Moyen Atlas Marocain' in *Mémoires et Documents, Nouvelle Série*, Vol. 4, pp. 283–294, Centre de Recherches et Documentation Cartographiques et Géographiques, CNRS, 1967.

MONOD T., 'L'Adrar Mauritanien', *Bulletin de la Direction des Mines*, A.O.F., No. 15, pp. 166–190, 1952.

OLLIER C. D. and HARROP J. F., 'The Caves of Mont Hoyo', *Bulletin of the National Speleological Society*, Vol. 25, pp. 73–78, 1963.

PETIT M., 'Contribution à l'Etude des Reliefs Grani-
tiques à Madagascar', *Association des Géographes de
Madagascar*, 1970.

RENAULT P., 'Le Karst du Kouilou', *Revue de
Géographie de Lyon*, Vol. 34, pp. 305–314, 1959.

VISSER D. J. L. 'The Geology of the Barberton Area',
*South Africa Geological Survey Special Publication*,
No. 15, 1956.

# CHAPTER SEVEN

# Plains and inselbergs

Across the face of Africa there are wide level plains, some of them stretching as far as the eye can see for hundreds of kilometres (Plate 7.1). Rising above the plains are occasional isolated hills called inselbergs, several of which are flat-topped, others like huge rounded domes and others in the form of piles of rock and boulders (Plate 7.2).

## The plains

Some of the plains have clearly been formed by the deposition of sediments, especially in some of the large structural basins like the Congo (Fig. 1.14).

Plate 7.2 Balancing Rocks, an inselberg of granite boulders at Epworth, 10 km SE of Salisbury, Rhodesia. The granite has been broken down by weathering along the major joints

Plate 7.1 The wide, shallow channel of the Niger river flowing across the flat savanna plains of West Africa, near Niamey

But a much larger proportion have been planed down to their present flat surface by denudation. These denudation plains or planation surfaces, as they are called, occur in all regions of the continent, but they are most impressive in the desert and savanna lands where the absence of thick vegetation clearly emphasises the planation.

Many of the plains have been eroded across a wide range of different rock types and structures, and except for a few residual hills are almost perfectly flat. Such a plain represents a late stage in the geomorphic cycle (page 11) and its formation involves a long period of geological time during which a cycle can run its full course without interruption. Any interruptions, such as changes of climate or alterations in base level due to earth movements, would automatically halt the progress of the cycle and initiate a new one. However, if no interruptions occur the landscape will be slowly reduced to a low-lying level plain.

For many years geomorphologists have worked towards an understanding of the origin and possible age of these plains. One way of dating a surface is from the age of deposits found on it. However, such deposits are not always present and the task of identification is therefore difficult. The most comprehensive study of the African planation surfaces is that by L. C. King. Fig. 7.1 is a summary of the main surfaces, according to age, based on King's classification.

When they were originally formed the plains were much nearer to sea level, but the intermittent uplift of the continent that initiated new cycles of denudation has meant that most surfaces now lie hundreds of metres above sea level. Older surfaces lie above younger and are often separated from them by a scarp or area of broken relief. The best example of such a scarp is the Great Escarpment of southern Africa, which reaches its highest in the Drakensberg on the Natal–Lesotho border (Plates 7.3 and 5.3). Above the scarp lie the remnants of the ancient Gondwana Surface. It is estimated that the scarp has retreated back from the east coast at an approximate rate of 1 metre for every 350 years (1:500,000 South Africa sheets SE 29/26, SE 31/28).

| Major Planation Surfaces in Africa | | |
|---|---|---|
| Name | Approx. age | Example Areas (metres a.s.l.) |
| Gondwana | Jurassic. (Before the break-up of the Gondwanaland continent.) | Drakensberg, Lesotho 3000–3300 m<br>Cherangani Mountains, Kenya 2700 m +<br>Nyika Plateau, Malaŵi 2100 m +<br>Jos Plateau, Nigeria 1400–1500 m |
| African | Cretaceous and Early Miocene. (First main surface developed in the 'new' African continent.) | Muchinga Mountains, Zambia 1350–1450 m<br>High Veld, South Africa 1000–1500 m<br>High Veld, Rhodesia 1200–1500 m<br>High Plains, Nigeria 600– 730 m |
| End-Tertiary | Late Miocene and Pliocene. (Covers a bigger area than any other surface in Africa.) | Zambian Plateau<br>Central Plateau, Tanzania<br>Northern Uganda<br>North-west Ghana |
| Quaternary | Pleistocene and Recent. (Several recent planations are included in this category.) | Mainly developed along coastal areas and in river valleys, e.g. valleys of Niger, Benue, Orange, Zambesi and Volta. |

Fig. 7.1 Major planation surfaces in Africa

Plate 7.3 The Amphitheatre, Drakensberg Escarpment, South Africa, above which the land rises to the high peak of Mont Aux Sources (3,299 m). Over the scarp face, the Tugela river (right foreground) drops 840 m in a series of five waterfalls

In most parts of Africa the very old surfaces have largely been destroyed by the denudation of later cycles and they now appear only as remnants on the tops of hills. Where such remnants are visible on a skyline they are known as accordant summits.

The height of the major surfaces above sea level varies from one part of the continent to another. This is due to the same differential upwarp and depression that produced the 'basins' and 'swells' of Africa (Fig. 1.14). In general, the height of the surfaces above sea level declines steadily from south to north, for southern and eastern Africa have experienced greater overall uplift than the remainder of the continent.

The existence of so many very marked planation surfaces has given Africa what may be called a polycyclic landscape. Polycyclic describes a landscape that has been subjected to a number of erosion cycles, each of which was interrupted in a late stage by the uplift which initiated a new cycle. A new cycle tends to advance most rapidly up the valleys of the major rivers, and the step between the new surface and the old is often marked by a waterfall (Plate 7.4 and 1 : 500,000 South Africa sheet 29/18).

Although these level plains with their inselbergs are acknowledged as being erosional in origin there is still considerable difference of opinion concerning the actual processes involved. In particular, it is not certain whether they were formed under the present climates, nor is it certain whether they were all formed in the same way. For example, many geomorphologists feel that the plains and inselbergs of the Sahara and Namib deserts may have originated under a past savanna climate, perhaps as long ago as the Tertiary era.

Plate 7.4 Augrabies Falls, South Africa, 130 km downstream from Upington, where the Orange river falls 145 m and leaves the End-Tertiary surface on its final stage to the Atlantic. As the falls retreat upstream so the Quaternary surface is extended inland

## The inselbergs

Inselberg is a word of German origin, meaning 'island-mountain'. It was first used in 1900 by B. W. Bornhardt to describe the remarkable dome-shaped hills of crystalline rock that he saw in Tanganyika. Later, S. Passarge described the inselbergs of South West Africa and introduced the idea of early writers that these features were a special type of residual hill found only in arid and semi-arid areas. This is not correct, for inselbergs are found in all climatic regions of Africa from arid to equatorial. They vary in shape and size, some being only small hillocks less than 100 m high while others, like the Brandberg of Namibia, rise over 1800 m above the plain.

Some inselbergs rise straight from the plain, but others are surrounded at their foot by a gently sloping rock platform called a pediment. A number of authors classify all residual hills as inselbergs including, for example, buttes and mesas (page 118). Other authors restrict the term to the steep-sided crystalline domes called bornhardts. In the present context inselbergs will be classified into two main types.

### Bornhardts and domes

A bornhardt is a high steep-sided dome-shaped hill, formed mainly in massive crystalline rocks, such as granite or gneiss. Less upstanding features with a similar rounded outline are usually described simply as domes, while low elongated domes are called whalebacks.

## Castle kopjes

A castle kopje is a steep-sided pile of massive crystalline boulders. Such hills often have a castle-like profile, hence their name. They are very similar to the tors of SW England, and in many African countries they are called tors. Some authorities consider tors and castle kopjes to be different landforms formed by different processes, but in the present context they will be taken as one. (Note: the word 'kopje' by itself is Afrikaans for hill, and in South Africa is often applied to any type of isolated hill.)

A number of inselbergs are undoubtedly composed of more resistant rocks than the surrounding plains, and for this reason obviously appear as upstanding residuals. But the majority are formed in rocks identical to the adjacent plains, and their origin would seem to be clearly related to the origin of the plains.

## The origin of plains and inselbergs

In the cycle of erosion, as originally proposed by W. M. Davis (page 11), the level plain to which the landscape is finally reduced was called a peneplain and the main processes involved in its formation were outlined in the concept of peneplanation. Because of their remarkably level nature the plains of Africa were once described as being peneplains. But the principles of peneplanation are modelled on the landscape of the humid temperate lands of Europe and eastern North America. In these areas the nature of weathering and erosion is very different from Tropical Africa, and, although Davis proposed a special cycle for desert regions, the concepts are not really acceptable in view of the processes that recent research has shown to be operating in the tropics. This research has led to the advancement of new theories to account for the origin of the landscape of inselbergs and plains. In particular there are two concepts of significance.

## Pediplanation

Pediplanation accounts for the origin of plains and inselbergs by the processes of scarp retreat and extension of the pediment. A plain is formed by the destruction of an older, higher plain along the scarp, the steep slope that separates the two plains (Fig. 7.2). At the scarp, where rivers fall from the original higher surface to the lower one, erosion is active and in conjunction with weathering and mass wasting this causes the gradual retreat of the scarp.

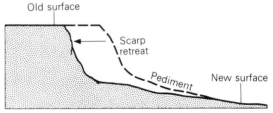

Fig. 7.2 Parallel scarp retreat and extension of the pediment

At the foot of the scarp a rock surface of low gradient, called the pediment, slopes gently down to the plain. On the pediment erosion is less active and sheet flooding (page 72) is the main process. This is sufficient to cause the extension of the pediment, a process called pedimentation, as the scarp retreats. The greater erosive activity on the scarp compared with the pediment helps maintain the knick, the sharp break of slope between the two.

Over the level surface of the plains erosion is limited and rivers tend to develop wide shallow channels (Plate 7.1). This is partly the result of the low gradients.

The overall effect is that the lower plain is enlarged at the expense of the higher plain. But there is no appreciable lowering of either surface (Fig. 7.3). The processes involve a wearing-back at the scarp face, rather than a wearing-down of the whole landscape as is the case in peneplanation.

As the process of scarp retreat goes on the pediments are gradually extended until all that remains between two encroaching pediments is a residual hill rising above the plains: an inselberg, the final remnant of an original higher planation surface. Initially, the inselbergs would be large and of the bornhardt variety, but eventually they are broken down into castle kopjes by weathering

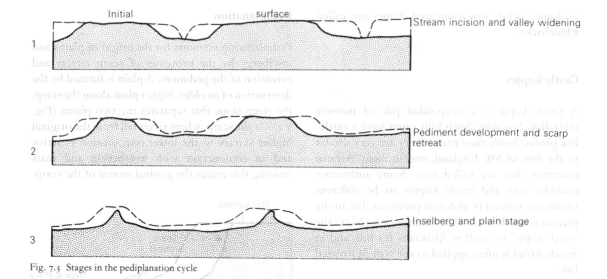

Fig. 7.3 Stages in the pediplanation cycle

attacking the joint planes. Finally, when the inselbergs are worn away, the pediments coalesce to form a continuous plain called a pediplain.

Thus, following the concept of pediplanation, the various planation surfaces of Africa may be described as ancient pediplains, one above the other, each of which is slowly being destroyed by the recession of the escarpment that divides it from the next surface below. If the surfaces are being destroyed in this way it explains why remnants of them still survive even though they were bevelled more than one hundred million years ago.

### Deep weathering and stripping

Many authorities now consider that pediplanation is limited to arid and semi-arid areas, and is not a satisfactory answer to the plains and inselbergs of the more humid forest and savanna lands. In savanna regions, it is argued, plains are more the result of deep weathering followed by stripping of the weathered layer and scarp retreat. The deep weathering by chemical decomposition, common in the humid tropics, may produce a layer of rotted rock up to 60 m thick, although this will vary from place to place according to the mineral composition and jointing of the rock (pages 63–5). In areas where joints are widely spaced and the rock is chemically

resistant the depth of weathering will be limited and fresh unweathered rock may even be exposed at the surface. The depth of weathering will therefore be unequal from place to place.

Streams flowing across the plains develop wide shallow channels and, due to both the low gradient and the fine nature of the chemically weathered load, vertical erosion is not great. However, if regional uplift initiates a new erosion cycle, streams will be rejuvenated and will cut down into the weathered layer. Where weathering was deep the material is easily stripped off by the streams, and a new lower plain is formed which is gradually extended by scarp retreat. But, in those areas where the rock was resistant inselbergs of hard unweathered rock will rise above the plain.

Thus, inselbergs are considered to evolve during two cycles of erosion, first a process of sub-surface differential weathering and second exhumation by stripping away the surrounding weathered material (Fig. 7.4A and B). It seems likely that several erosion cycles are involved in the formation of very large bornhardt inselbergs exceeding 100 m in height, since it is doubtful if weathering could be effective to such great depth during a single cycle. The Namuli Mountains (2124 m) in northern Mozambique is an example of a group of very large bornhardts. These peaks, which rise 1000 m above the surrounding landscape, lie SW of Entre Rios.

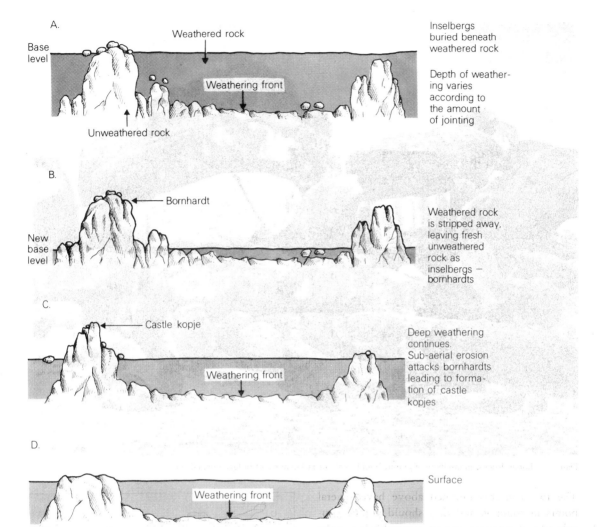

Fig. 7.4 The evolution of inselbergs (A, B, C, D)

The actual form an inselberg will take depends on the spacing of the joints. Wide-spaced joints produce bornhardts, and close-spaced joints produce castle kopjes. In rock where very close jointing leads to almost complete breakdown, the effect of stripping will be to reveal a pile of rounded boulders or corestones (Plate 7.5).

Once bornhardts are exposed, surface weathering and erosion begin to attack the curved joint systems leading to gradual exfoliation (Plate 4.4). Bornhardts with well-developed sub-rectangular jointing may suffer extensive attack and collapse into castle kopjes (Fig. 7.4C). Thus castle kopjes are thought to evolve both directly by deep weathering and indirectly by the collapse of bornhardts. However, not all bornhardts are broken down in this way and some particularly massive ones may survive through several successive erosion cycles. Eventually, if the cycle of erosion is not interrupted by further uplift, a plain may be cut at base level in which all the inselbergs are planed down so that fresh rock appears only as an extension of the general level of weathered rock (Fig. 7.4D).

Plate 7.5    Bongo Rocks, an inselberg of granite boulders about 15 km north of Bolgatanga, Ghana

The two concepts outlined above have several points in common and they should not be considered as alternatives, since it is quite likely that the African plains have originated in more than one way. In many regions, such as Nigeria, there is strong evidence to show that inselbergs have been shaped by sub-surface weathering, while in southern Africa and eastern Kenya the evidence points to pediplanation.

### Examples

*The plains of northern Nigeria*
In northern Nigeria two main planation surfaces can be identified: the Gondwana and the African (Fig. 7.5). The African is the most extensive and, lying between about 450 m and 730 m, it forms what

Fig. 7.5  Major planation surfaces in northern Nigeria

are generally known as the High Plains. The plains cover a large part of northern Nigeria, extending from Bauchi in the east almost to Sokoto in the west. It is a vast region of negligible relief across which meandering streams flow in broad shallow valleys. Near Zaria the River Galma, for example, a tributary of the Kaduna, has an average gradient of only about one-third of a metre per kilometre.

In certain areas the African surface is divided by a low scarp or zone of broken country into two levels, the upper of which lies above 600 m. The plains are developed across igneous and metamorphic rocks of the basement complex. But unweathered rock rarely outcrops, except in the castle kopje and bornhardt inselbergs that rise from the plains. The weathered layer is mostly topped by laterite (page 68), and water boreholes have shown that the total depth of weathering often exceeds 50 m. Where the laterite cover is broken erosion has sometimes formed small mesas with lateritic caps.

The Gondwana planation is preserved at about 1400 m on the Jos Plateau, and it is one of the few remnants of this surface in the continent north of the equator. The African surface ends abruptly at the foot of the scarp that surrounds the Jos Plateau (Plate 7.6). The plateau owes its continuing existence largely to the resistance of the Younger Granite ring complexes (pages 58–9). The scarp that separates the Gondwana surface of the plateau from the African surface of the High Plains is particularly steep and high on its south side, but more gentle to the east and north. In spite of its resistance the plateau is gradually being destroyed by the retreat of the scarp, which in some areas now lies nearly a kilometre from the original boundary of the Younger Granites.

In contrast to the Jos Plateau scarp, the scarp that separates the African surface of the High Plains from the lower surrounding End-Tertiary surface is often indeterminate. In the NW and NE the End-

Plate 7.6 The steep southern scarp of the Jos Plateau separating the higher Gondwana surface from the lower African surface

Plate 7.7 Kufena Hill, a huge granite inselberg that rises from the High Plains west of Zaria, Nigeria

Tertiary gradually changes into the depositional plains of the Sokoto and Chad Basins, while southwards it is being destroyed by the encroaching Quaternary surface along the scarps that border the Niger and Benue Valleys.

The surface of both the Jos Plateau and the High Plains is interrupted at intervals by steep-sided granite and granite-gneiss inselbergs. Some of these are true castle kopjes, for example north of Koriga (Nigeria 1:100,000 sheet 122) and near the Zaria–Kaduna road (1:50,000 sheet 124NW). Others are in the form of domes divided into blocks by a system of vertical and horizontal joints (Plate 7.8). But by far the largest of the inselbergs are the giant bornhardts, such as those near Kusheriki (1:100,000 sheets 121 & 142), Abuja and Zaria. Dutsin Zuma, 8 km SW of Abuja, is a huge dome with precipitous sides that rises over 300 m above the plain (1:100,000 sheet 186). Kufena Hill dominates the landscape west of Zaria (Plate 7.7). It is a giant elongated bornhardt, lying 5 km from the town, and consisting of three overlapping domes partly separated from each other by deep joint-

controlled clefts (1:50,000 sheet 102 SW). The east side of Kufena is particularly steep while all the lower slopes are masked by talus formed from the exfoliation of the inselberg. Surrounding Kufena are very low domes, barely breaking the surface and sometimes topped by groups of rounded boulders or corestones. Most of the large inselbergs, like Kufena and Dutsin Zuma, are built of the hard older granite of the Basement Complex.

### Inselbergs in southern Malawi

In Malawi west of the Shire River both the African and the End-Tertiary planation surfaces can be identified. The African surface lies at about 1350 m to 1550 m and dominates the landscape from Dedza south towards Neno, near where there is a small east–west scarp. South of the scarp occurs the lower End-Tertiary surface sloping from about 1050 m to 600 m.

Both the planation surfaces are largely eroded in Precambrian gneiss, and rising above the gently sloping bush-covered grasslands are numerous inselbergs, especially in the border region north of

Plate 7.8 One of a number of partly collapsed bornhardts that rise from the surface of the Jos Plateau. The inselberg is being gradually broken down into cuboidal blocks by a system of intersecting joints

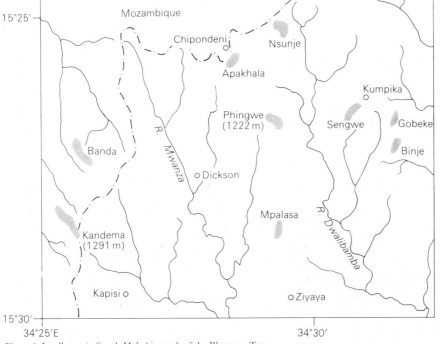

Average height of the plains is 750–1050 m

Fig. 7.6 Inselbergs in South Malaŵi, north of the Blantyre–Tete road

Plate 7.9 Bismarck Rock, at Mwanza, Tanzania. A partly drowned granite castle kopje inselberg on the south shore of Lake Victoria

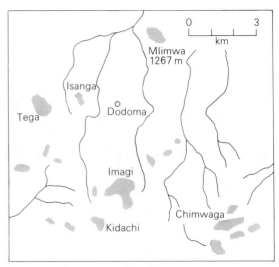

Fig. 7.7 Inselbergs near Dodoma, Tanzania

Mwanza (Malaŵi 1:50,000 sheets 1534 A4, 1534 B3). Here there is a variety of exfoliating domes and bornhardts. Phingwe (Fig. 7.6) is a perfect example of a huge rounded inselberg, and while some, like Nsunje, have a tilted profile, due to the alignment of the main joints, others, like Banda, are in the form of elongated whaleback ridges.

*Inselbergs near Dodoma, Tanzania*

Large numbers of inselbergs, both bornhardts and castle kopjes, rise above the End-Tertiary surface of the central plateau area in Tanzania. Much of this region is underlain by a huge granite batholith and most of the inselbergs are formed in granite. They vary in size from smallish castle kopjes, such as Bismarck Rock at Mwanza (Plate 7.9), to giant bornhardts like Uhambingetu (Wambengetu) at about 30 km NE of Iringa which rises more than 450 m above the plain.

Near Dodoma (Fig. 7.7) the inselbergs are more moderate in size and are often composite in form (Tanzania 1:50,000 sheets 162/1, 162/2). They tend to have a dome base, but are often topped by cuboid blocks, producing a castellated profile. One such is Lugala Rock, about 20 km NW of Dodoma, which consists of two monolithic blocks rising from a small dome covered with granitic boulders. Several inselbergs also surround the town, notably Mlimwa, Tega and Imagi. They have a domed shape

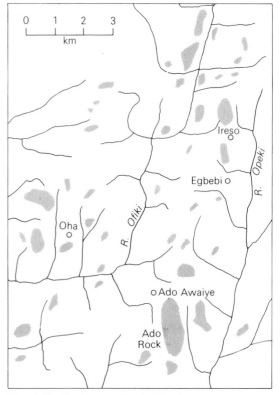

Fig. 7.8 Bornhardts west of Iseyin, Nigeria

148

and are generally elongated in a NW–SE direction. The plateau here is about 1000 m to 1200 m above sea level. Mlimwa (Lion Rock) is the most significant of the inselbergs, and is composed of a large steep-sided monolith rising some 30 m above a whaleback base disguised by scrub and boulders.

*Bornhardts and domes near Iseyin, Nigeria*
North of Abeokuta and west of Iseyin in the grass plains of Oyo State a long line of granite and gneiss inselbergs, of the bornhardt variety, extend north–south approximately between Shaki and Eruwa (Plate 7.10). They are sometimes known collectively as the Ogun Hills (Nigeria 1:100,000 sheets 220 & 240). The highest and most outstanding of these features is Ado Rock (Fig. 7.8), but other prominent ones include those at Oke Iho, Oke Ile, Shaki, and Eruwa. Most of the inselbergs rise many metres above the End-Tertiary surface of the plains, which are at a general level of 200 m, and they are often arranged together in groups. Some, like those at Oke Iho, surround a central tableland area, and in the past nearly all the large rocks

provided important defence sites. Between the bornhardts, areas of bare rock with a slight domed profile occasionally break the surface. These have been described as 'ruwares' and they are thought to be either the peak of an emergent dome or else the remnant of a collapsed bornhardt.

Ado Rock (Plate 7.11) lies south of the village of Ado Awaiye on the Iseyin–Eruwa road. The bornhardt rises nearly 450 m above sea level and its south face is almost a sheer cliff of 200 m. On its summit area are two rock basins containing small lakes which show that the inselberg includes no major joints or else the lakes would drain away.

Other examples of bornhardts are in the forest belt of southern Ghana between Nsawam and Cape Coast, e.g. Nyanoa near Nsawam and Abura Abura near Fanti Nyankumasi; near Sinda in eastern Zambia, e.g. Mbewa and Simvwa; near Salisbury and Mtoko in Rhodesia, e.g. Domboshawa 30 km north of Salisbury; and in the central Namib Desert, e.g. Klein Spitzkop (Plate 8.11). Examples of castle kopjes occur near Funsi, 60 km NE of Wa, northern Ghana; near Marandellas, 65 km SE of Salisbury,

Plate 7.10 Domed inselbergs rising from the plains west of Iseyin in Oyo State of Nigeria

Plate 7.11 The massive bornhardt inselberg of Ado Rock, 70 km north of Abeokuta, Nigeria

Rhodesia; in the Accra Plains, SE Ghana (e.g. Cherekecherete); and in the Maragoli Hills, 20 km NE of Kisumu, Kenya.

*A classification of mountains*

Three major mountain types have already been considered in Chapters 2 and 3. The very large inselbergs described in this chapter form a fourth type, known as residual mountains. A simple classification of mountains, according to origin, is given in the following chart.

| | Examples: |
|---|---|
| A. Block Mountain | Ruwenzori, page 20 |
| | Usambara Mountains |
| B. Fold Mountain | Atlas Mountains, page 29 |
| | Cape Ranges, page 29 |
| C. Volcanic Mountain | Kilimanjaro, page 46 |
| | Mount Cameroon, page 46 |
| D. Residual Mountain | Namuli Mountains, page 142 |
| | Hombori Mountains, page 119 |

## References

COTTON C. A., 'The Theory of Savanna Planation', *Geography*, Vol. 46, pp. 89–101, 1961.

COTTON C. A., 'Plains and Inselbergs in the Humid Tropics', *Transactions of the Royal Society of New Zealand*, Vol. 88, pp. 269–277, 1962.

GIARDINO J. R., 'The Inselberg Landscape of Eastern Province, Zambia', *Zambia Geographical Association Magazine*, No. 23, 1973.

HARPUM J. R., 'Evolution of Granite Scenery in Tanganyika', *Records of the Geological Survey, Tanganyika*, Vol. 10, 1963.

HILTON T. E., 'The Landforms of North Western Ghana', *Journal of the West African Science Assoc.*, Vol. 11, pp. 136–149, 1966.

HILTON T. E., 'The Accra Plains: Landforms of a Coastal Savanna of Ghana', *Zeitschrift für Geomorphologie*, Vol. 10, pp. 369–386, 1966.

JEJE L. K., 'Inselberg's Evolution in a Humid Tropical Environment: the example of SW Nigeria', *Zeitschrift für Geomorphologie*, NF, Vol. 17, pp. 194–225, 1973.

KING L. C., 'A Theory of Bornhardts', *Geographical Journal*, Vol. 112, pp. 83–87, 1948.

KING L. C., *The Morphology of the Earth*, 2nd Ed., Chaps. 5 and 9, Oliver and Boyd, 1967.

MORTIMORE M. J., 'Zaria and its Region', Geography Dept., Ahmadu Bello University, *Occasional Paper 4*, 1970.

OJANY F. J., 'The Inselbergs of Eastern Kenya, with special reference to the Ukambani area', *Zeitschrift für Geomorphologie*, Vol. 13, pp. 196–206, 1969.

OLLIER C. D., 'The Inselbergs of Uganda', *Zeitschrift für Geomorphologie*, Vol. 4, pp. 43–52, 1960.

THOMAS M. F., *Tropical Geomorphology*, Macmillan, 1974.

# CHAPTER EIGHT

# Desert scenery

Africa, as a result of its shape and latitudinal position, has the misfortune of owning two large areas of desert and semi-desert, one of which is the greatest tropical desert in the world, the Sahara.

The Sahara is over $7\frac{1}{2}$ million sq km, that is one quarter the size of Africa. It stretches for about 5500 km from the Atlantic to the Red Sea, and in some places its north–south extent exceeds 1500 km.

In southern Africa the Kalahari covers more than 500,000 sq km. It is separated from the coastal Namib Desert by the mountains of Namibia.

The deserts of Africa owe their origin to meteorological causes and there is nothing inherently barren in their geological structure. Geological, biological and archaeological evidence all point to a much wetter climate in the recent past (Plate 1.7).

The essential characteristic of the desert is its aridity, and this is mainly the result of low rainfall and high evaporation. Deserts can be defined as areas receiving less than 250 mm of rain a year, although this is only an approximation since the mountains of the Sahara receive more than this while much of the Kalahari has over 500 mm. Since the desert has been defined by climatic factors it is to be expected that within its frontiers there will be a considerable variety of rock type and landform. Nowhere is this more apparent than in the Sahara.

## The Sahara

The relief of the Sahara ranges from the high volcanic mountains of the Tibesti, 3415 m, and the Hoggar, 2918 m (Plate 8.1), to the low lying depression of the Qattara, 134 m below sea level. Of intermediate height are gigantic plateaus and uplands bordered by steep escarpments and cut by

Fig. 8.1 The Sahara, showing the location of major features described in Chapter 8

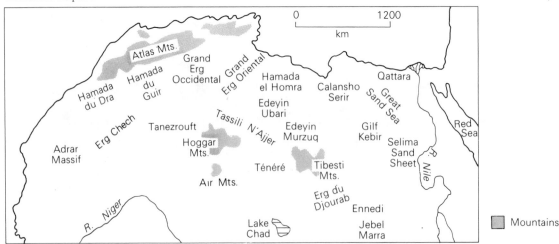

Plate 8.1 The volcanic peaks of the Atakor, the rugged mountainous district that forms the central part of the Hoggar

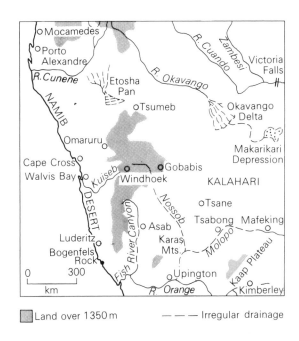

Land over 1350 m    — — — Irregular drainage

Fig. 8.2 The Kalahari and Namib Deserts

deep valleys, such as Gilf Kebir in Egypt (pages 165–66) and the Mauritanian Adrar.

But much of the Sahara is dominated by vast sandy or stony areas of level monotonous surface, for example the Tanezrouft (Plate 8.2) and the Tenéré. The Tenéré, a huge expanse of flat gravel and sand between the Air Massif and the Tibesti, is one of the most inhospitable areas on earth. Throughout the Sahara there is an almost total lack of any regular co-ordinated surface drainage. The main exception is, of course, the Nile which flows north for over 1500 km without a single tributary.

Possibly the most distinctive characteristic of this arid environment is the sharp angular nature of the landscape. The rounded slopes and green colour of the humid tropics are replaced by steep rugged cliffs and fantastically carved bare rock outcrops in vivid reds, yellows and browns.

The bare, rocky desert where boulders and residual hills separate large areas of broken rock is called 'hamada'. Examples include Hamada el Homra in northern Libya and Hamada du Guir on

Plate 8.2 The level stony surface of the barren Tanezrouft, west of the Hoggar Mountains, Algeria

Plate 8.3 Transverse dunes in the Edeyin Ubari, Western Libya

the Algeria–Morocco border. This rock desert contrasts with the flat, stony plains called 'regs', like the Tanezrouft (Plate 8.2). In Libya these gravel-covered plains are called 'serir', such as Calansho Serir. The sandy desert, including both small dune fields and huge sand seas, is known as 'erg' or 'edeyin' (Plate 8.3). Many ergs are several thousand square kilometres in area, although sandy desert covers less than 15% of the Sahara. The Grand Erg Oriental of Algeria is larger than either Malawi or Sierra Leone.

### The Kalahari and the Namib

In contrast to the Sahara, the Kalahari is smaller and has a higher rainfall causing some of it to be covered with grass and scattered trees. It is a semi-arid region rather than a true desert. The Kalahari is a vast level plain with little upstanding relief. It stretches from the Orange River in the south to the Okavango Delta and Etosha Pan in the north. Nearly all of its area is covered by great thicknesses of red Kalahari sand, deposited during the Tertiary

Plate 8.4 Chemical and physical weathering have caused exfoliation on the face of these rocks, near the rail line, south of F'Derik (Fort Gouroud), Mauritania

and Quaternary eras in one of the great structural basins of Africa (Fig. 1.14). It is probably the largest area of continuous sand-covered desert in the world.

The Namib Desert extends from Moçamedes in the north to the Orange River in the south, and is one of the most desolate of coasts. It is halted inland by the discontinuous line of the Great Escarpment, and is 160 km at its widest near Luderitz. Much of it is covered by sand dunes, but in the south there are extensive gravel plains similar to the reg of the Sahara, and in the north large outcrops of bare rock.

## Agents of earth sculpture in desert areas

The major desert landforms are, as in other environments, the result of either erosion or deposition. The agents involved in the origin of these landforms are weathering, water and wind. However, the climate and the general absence of vegetation

mean that the relative roles of these agents are not the same as in the more humid parts of Africa.

### Weathering

The thick layers of rotted rock found in many parts of the deserts seem to provide evidence of active weathering (Plate 8.4). However, it is important to remember that not only is weathering very slow in arid areas but also that some of the great spreads of weathered rock may be a relict from a more humid past climate (see page 67).

### Wind

Wind is a far more significant agent of transport and deposition than of erosion. The shapes of the dunes in the vast Saharan ergs are all the work of wind. Wind transports sand and other fine particles mainly in suspension, and by saltation and surface creep.

Plate 8.5 Weathering and wind erosion have combined to produce this giant mushroom-shaped rock pedestal, 'The Devil's Rock', north of Agadez, Niger. A similar example is the 30 m tall Mukarob (Finger of God) near Asab (Fig. 8.2), Namibia

Saltation is the most important movement and is a kind of jumping action caused by turbulence of the wind.

Wind erosion involves two main processes, deflation and abrasion, of which the former is the more potent. However, few major features are due to wind erosion alone, and in those landforms where wind plays a role, for example yardangs, rock pedestals and blowouts, the rocks have generally been already weakened by weathering (Plate 8.5).

Deflation is the removal of soft loose materials, such as weathered debris, by a lifting and rolling action. Abrasion is the blasting action of wind-blown sand. During sand storms abrasion may be very effective, polishing hard rocks and under-cutting weak rocks, but since wind can lift few materials above 50 cm most abrasion is at ground level. Perhaps the main effect of abrasion is the production of ventifacts, pebbles polished and faceted by the wind. Vast sheets of ventifacts cover the regs of the Sahara and the stony parts of the northern Namib.

## Water

In spite of the low humidity and low rainfall of desert areas, water is nevertheless the dominant agent of erosion and deposition. Even the very dry parts of the eastern Sahara record some rain, although it may be only once in ten years. It is important here to differentiate between the very arid desert areas, such as southern Libya, and the semi-arid desert margins. In the former weathering and water erosion may be limited, but in the latter, on the border of the savanna zone, surface run-off is very powerful in the short wet season.

The action of run-off is aided by the scarcity or absence of vegetation. Sheetfloods (page 72) develop on the pediments and gently sloping surfaces that surround uplands, while in steep-sided valleys raging torrents carry away enormous loads of weathered debris, eventually to be deposited in low lying depressions or as alluvial cones at the edge of uplands. In a fairly short time, however, the water percolates into the ground or is lost by evaporation.

Throughout the deserts, drainage is short-lived and intermittent. Channels may have water flowing in one part but not another and only in a few areas is there any semblance of an organised drainage network. In some regions, such as the Hoggar Mountains, an integrated system of dry valleys can be seen and these have clearly been carved by more regular surface run-off under a wetter climate of the past. Many desert landforms, in fact, may be relics of a past humid climate (pages 168–70).

## Desert landforms

### Yardangs

These are elongated rock ridges, generally aligned with the prevailing wind. They are usually less than 10 m high and 500 m long, but some are more. Most are cut in soft rocks along lines of weakness by wind erosion. The upstanding ridges or yardangs are sometimes carved into fantastic shapes and are usually undercut on their windward sides. If the rocks are composed of hard and soft bands, and differential erosion produces ridges with a resistant cap the features are called zeugen.

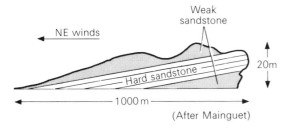

Fig. 8.3 Diagrammatic section through one of a number of very large yardangs cut in sandstone near Ounianga Kebir, NE Chad

*Yardangs in Egypt*

East of the Nile, in the Silsila Gap, 10 km north of Kom Ombo (Fig. 8.4), a large group of yardangs has been eroded from Pleistocene alluvial deposits. The elongated ridges are up to 10 m high and mainly have a NW alignment. Similar formations have been carved from old lake sediments in a depression near Abu Ballas (Fig. 8.4 and Plate 8.6), along the route from Dakhla Oasis to Gilf Kebir.

Other examples are: near Ounianga Kebir (Fig.

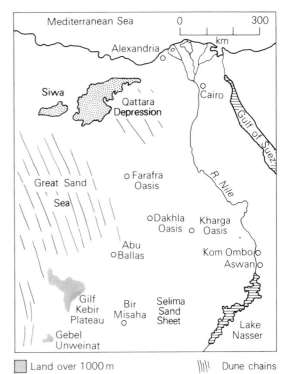

Fig. 8.4 The Western Desert, Egypt

8.3), 180 km NE of Faya Largeau, Chad (Fig. 8.9), NE–SW alignment; near Tajarhi, 60 km south of Al Qatrun (El Gatrun), SW Libya (Fig. 8.15), NE–SW alignment; and near In Salah, central Algeria (Fig. 8.15).

Plate 8.6 Yardangs eroded from mud deposits NE of Gilf Kebir in Egypt. They are undercut on their windward sides

## Blowouts (deflation hollows, pans)

These are shallow depressions in outcrops of weak rock. Some are several square kilometres in size, but most are small. They are deepened by wind deflation, but there is doubt as to whether the initial depression is always due to wind, and some may originate in faulted rocks. The hollows retain moisture more than adjacent areas, and this encourages chemical weathering on the floor making for easier removal of material by the wind. The depth is limited by the water table, since once this is reached wind will not easily be able to lift out wet particles. In some cases, when the water table is reached, an oasis lake may form on the floor of the depression (Fig. 8.5).

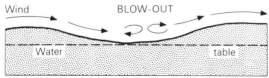
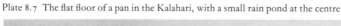

Fig. 8.5  Blowout

### Deflation pans in the Kalahari

In the Kalahari there are several thousand flat-floored depressions called pans. Most are quite small, but some are over 10 km wide and more than 30 m deep. They are particularly numerous near Tsane in Botswana (Fig. 8.6), and in the 'mier' country north-west of Upington, South Africa. The pans are generally dry, but after rain they may contain water for several months (Plate 8.7). Many have a grass cover, while others have a salt crust on their floor. Most pans are the result of weathering and deflation, and are formed in unconsolidated clays, shales and sands. Several seem to have developed along ancient stream channels, suggesting that some initial depression may be necessary before deflationary action can be effective. Further, a high proportion occur in areas where there is a thick layer of calcrete (page 69) at the surface, for example around Khuis near the Molopo River. These pans are often deep with steep sides, and calcrete solution is a major factor in their initiation.

In north Namibia the huge Etosha Pan, one of the largest salt pans in the world, occupies the lowest part of a large structural basin.

Plate 8.7  The flat floor of a pan in the Kalahari, with a small rain pond at the centre

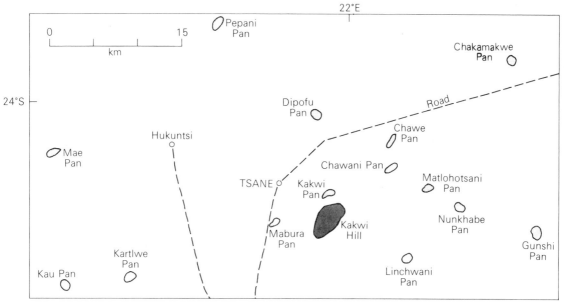

Fig. 8.6 Pans near Tsane, Botswana

## Sand formations

Most areas of sandy desert have a surface pattern of waves which range from small ripples to huge undulations. The size and shape of these sand waves depends mainly on the direction and strength of the wind and on the supply of sand. There are, according to size, three main groups of sand formation.

1 *Ripples*: these are the smallest, sometimes less than a centimetre high. They are mostly found on and between larger features called dunes.

2 *Dunes*: these are of intermediate size, although some rise above 100 m. Many form on the backs of very large formations called draas. There are several types of dune, but the main ones are:

a) barchans: crescent-shaped dunes with two horns pointing down wind. They may be up to 25 m high and 400 m wide, and occur in small groups or large ergs. Barchans form from small sand mounds by winds of moderate strength blowing from one direction only. Sand blown onto the mound is moved towards the crest and tends to accumulate in

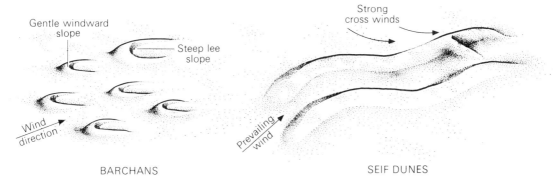

BARCHANS

SEIF DUNES

Fig. 8.7 Barchans (a) and Seifs (b)

the lee of the crest away from the full wind force. But sand concentrated here produces a steepening of the lee slope, which causes sand to slip down and so the dune slowly advances. Wind eddying also helps maintain the steep slip slope. The horns develop because the edges of the dune receive less sand and therefore move forward quicker (Fig. 8.7a).

b) seifs: longitudinal, steep-sided ridges, with knife-like crests rising to over 100 m and a length sometimes exceeding 100 km. They lie parallel to each other in vast complexes, and the flat inter-dune corridors vary in width from 25 m to 500 m. Seifs build up gradually from small sand ridges and, although aligned parallel with the prevailing wind, strong cross winds are the main factor determining their height and width. Where cross winds are always from one direction the dunes develop a gentle windward slope and a steep lee slope (Fig. 8.7b).

c) transverse dunes: long, wave-like ridges separated from each other by flat-bottomed troughs. They tend to form in huge dune fields, but most are less than 50 m high. They are built by light to moderate winds, blowing from one direction only and concentrating larger sand grains into a series of transverse ridges. In the western Sahara they are known as aklé (Fig. 8.8).

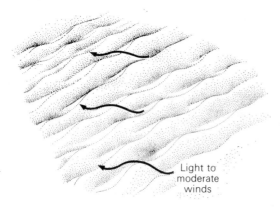

Light to moderate winds

TRANSVERSE DUNES

Fig. 8.8 Transverse dunes

3 *Draas*: these are the largest sand features, but like the dunes that form on them they are generally spaced at regular intervals. Many draas are similar in shape to seifs and transverse dunes. In some of the Saharan sand seas, notably the Great Eastern Erg, enormous star-shaped sand mountains or rhourds have formed at the intersections of draa ridges. These rhourds have huge radiating buttresses and often reach heights of 200 m.

### Seif dunes in the Egyptian Desert

The Great Sand Sea of Egypt and Libya extends south from the Siwa Oasis towards the Gilf Kebir Plateau, a distance of about 700 km (Fig. 8.1). The Sand Sea is one of 20 or more large Saharan ergs, many of which cover the underlying solid rock to depths of over 300 m. Although it is clear that wind has shaped and moulded the sand in these ergs into the present dune patterns, it is probable that the sand itself was originally deposited by water. Nearly all the ergs lie in large basin areas and are almost certainly the result of millions of years of deposition in lakes and shallow seas.

The main dune formation in the Great Sand Sea are seifs (World 1:1,000,000 Map, sheet NG 35). Some are over 100 km long and up to 100 m high. The inter-dune corridors between them are sometimes over 2 km wide, but usually less, and covered by undulating sand deposits. The dunes follow a SSE direction aligned with the prevailing winds. In the north of the Sand Sea the seifs lie on top of much larger sand formations called draas.

Other examples of Saharan seif dunes occur in Erg Chech, western Algeria (Fig. 8.1) and Grand Erg de Bilma, eastern Niger.

### Namib Sand Dunes

Much of the south and central part of the Namib Desert is dune-covered, but the area of greatest concentration is that between Walvis Bay and Luderitz (World Aeronautical Chart 1:1,000,000 sheets 3273 & 3302). Here, for 500 km, is an unbroken sea of huge longitudinal dune chains (Plates 8.8 and 8.9). These seifs extend from the coast inland for over 100 km and have an average height of 50 to 100 m, although several rise over 250 m. The dune crests have a constantly repeating S-shape, rising and falling to form a chain of summits

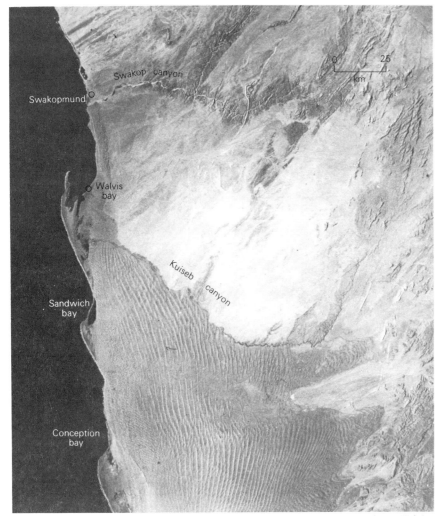

Plate 8.8 Satellite view of part of the central Namib desert, showing the great sea of seif dunes. Walvis Bay is sited in the lee of the northernmost of the spits

Plate 8.9 Seif dune crests near the Luderitz rail line

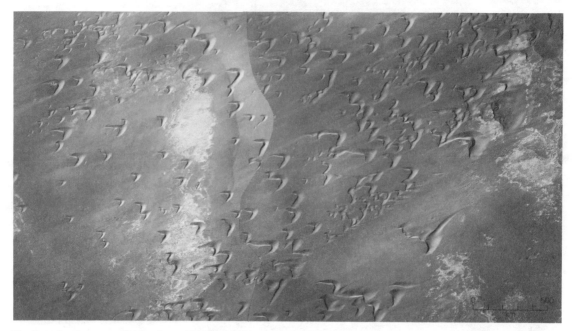

Plate 8.10 South west trending barchans of the Erg du Djourab, south of Faya Largeau, Chad

(Plate 8.9). The most important wind in their formation comes from the SW, which gives them gentle SW slopes of less than 15° and steep NE slopes exceeding 30°. However, occasional strong east winds are also important, which tends to destroy the symmetry of the dune pattern. Near the coast the dunes are white to yellow in colour, but inland where they are older the effects of oxidation have made them red.

Crossing the desert from east to west are several deep canyons, along which there are occasional floods. The Kuiseb and Swakop Canyons have a particularly marked effect on the dunes. In Plate 8.10, the Kuiseb crosses the centre of the photo in a NW direction, and it is clearly seen how its deep valley abruptly halts the northward advance of the dunes. North of the Kuiseb the dunes are limited to a narrow coast strip of only 10 km wide. They are finally stopped by the Swakop canyon some 65 km north of the Kuiseb.

In the Sahara examples of transverse dunes occur in the Edeyin Ubari, Libya (Plate 8.3).

*Barchans of the Erg du Djourab, Chad*
Although barchans occur in most deserts they are less common in the Sahara than, for example, in Arabia. One extensive field is the Erg du Djourab, south of the Tibesti Mountains in the Borkou region of Chad (1 : 200,000 IGN sheets NE-34-I, NE-34-II, NE-34-VIII). The erg lies in the northern part of the Bodélé depression, lowest area of the Chad Basin (Fig. 1.14). On average the dunes are less than 20 m high, but some are as much as 50 m. Most of the barchans are about 100 m wide from

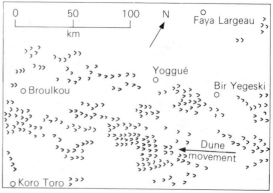

Fig. 8.9 Barchans of the Erg du Djourab, Chad

horn to horn, with a length of 150 m. Some are isolated and rise abruptly above the flat sandy plains, while others have coalesced so that their summits merge (Plate 8.10). The gentle windward slopes, often less than 5°, are in sharp contrast to the steep crescent fronts.

The barchans are mobile and move SW under the influence of the NE winds (Fig. 8.9). Measurements made near Faya Largeau have recorded movements of up to 25 m in a year. The dunes move intermittently, rather than continuously, and the period of greatest movement is usually between November and April when wind speeds are at their maximum.

Other examples of barchans in the Sahara are: in the Mourdi Depression, eastern Chad; the Tidi Dunes, one of a number of small dune fields east of the Djado Plateau, northern Niger; and several barchan groups which occur in Mauritania, for example west of Chemchane Sebkha in the south of the Erg Makteir (Fig. 8.13), and near the rail line, 50 km north of Nouadhibou.

## Desert plains

There are vast horizontal plains, hundreds and even thousands of square kilometres in area. Some are overlain by gravels, others by a layer of sand, while others are bare and reveal the eroded bed-rock.

Many seem to be the result of erosion by running water under an earlier, more humid climate. But some are scarred by deep grooves parallel to the prevailing winds. However, to date there has been insufficient research to explain the cause of all the plains.

### The Selima Sand Sheet

The Selima Sand Sheet (World 1:1,000,000 Map, sheet NF 35) is a vast desert plain, eroded from Nubian Sandstone on the Egypt–Sudan border. It slopes gently northwards, covering more than 40,000 sq km, between Gilf Kebir and Lake Nasser (Fig. 8.4). There are occasional dune fields and isolated residuals, but for the most part the plain is almost perfectly flat, and covered with a thin layer of sand less than 50 cm thick. The extremely level nature of the plain suggests it is probably a planation surface (Chapter 7), perhaps of late Tertiary age (Fig. 7.1).

Other examples of desert plains are: Ténéré, between the Tibesti and Aïr Mountains; Serir Tibesti, north of the Tibesti Mountains; and the Tanezrouft, west of the Hoggar Mountains (Plate 8.2).

## Inselbergs and other residuals

Many residual hills and rock masses rise abruptly from the desert floor. They vary greatly in size,

Plate 8.11 The granite inselberg group of Klein Spitzkopje, near Usakos, on the Windhoek–Walvis Bay road, Namibia. At their base the bornhardts are surrounded by a gently-sloping pediment

shape and origin, and their appearance is made the more dramatic by the absence of vegetation. Some are isolated, others occur in groups. In the Sahara many are called gara (plural gour), regardless of their nature. A common residual is the rock pedestal, shaped by wind abrasion and weathering undercutting and wearing away the weaker strata (Plate 8.5). Much larger features are inselbergs and buttes (Chapters 6 and 7). Buttes are often flat-topped outliers to a sedimentary plateau (Plate 6.2). Crystalline rocks tend to form inselbergs of the bornhardt (Plate 8.11) and castle kopje type. Other residuals are the vertically sided rock towers, generally carved from sedimentary outcrops (Plate 8.12). At the foot of many residuals there is frequently a rock pediment (page 140), often obscured by debris, sloping away to the surrounding plain.

Plate 8.12 Sandstone rock towers on the west of the Tibesti mountains, near Zouar, northern Chad

## Wadi

This is a deep, steep-sided desert valley, with rocky cliff-like walls that rise sharply from a flat floor (Fig. 8.10). Few wadis hold permanent streams, but they are usually flooded after a rain storm. When water does flow it tends to occupy a variable channel in a wide bed. Large quantities of weathered debris may collect in a wadi, with the result that when floods occur the stream has an enormous load which it is unable to move far before it develops into a mudflow (page 70), or evaporates or percolates into the ground. Streams tend to deposit their loads as alluvial fans where the wadi opens out onto a desert plain or inland drainage basin. The extensive network of wadis that dissect many of the Saharan uplands could not have formed in the present dry climate with its sporadic rainfall, and they are clearly a legacy of one or more past pluvial periods.

*Gilf Kebir Plateau, SW Egypt*

Gilf Kebir is a huge flat-topped sandstone plateau that rises more than 300 m above the desert floor, bordered by the Selima Sand Sheet to the east and the Great Sand Sea to the north (World 1:500,000 Map, sheet 568/8). The bounding cliffs are almost vertical in places. The height of the plateau declines northwards until it merges into the plains, while cut back into its edge are many steep-sided wadis (Fig. 8.11).

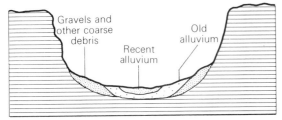

Fig. 8.10 Wadi cross-section

The plateau top is a good example of a remnant planation surface (Chapter 7). According to L. C. King, its flat top is part of the African Surface and contrasts with the lower End-Tertiary plain to the east, now largely covered by the Selima Sand Sheet.

Fronting the plateau scarp are numerous residuals of varying sizes, some flat-topped buttes, others much smaller and conical in shape, but all evidence of gradual scarp retreat. The rate of denudation now is considerably less than in the past when rainfall was higher. Present-day processes are chiefly weathering and mass wasting, combined with occasional downwash. The Nubian Sandstone of the plateau is porous and denudation may also result from rainwater percolating downwards and sapping the cliff foot where it reappears.

In his study of Gilf Kebir, R. F. Peel has given a clear description of the wadis that dissect the plateau:

Each wadi runs in a chain of straight lines connected by abrupt bends; the smooth curves of a normal river valley are conspicuously absent. Tributaries join them at all angles. In cross-section they are typically flat-floored, the surface rising only gently to the base of the cliff walls, except where sand drifts have accumulated. The cliff walls are invariably steep, showing an aver-

Land over 600 m

Land over 800 m

Fig. 8.11 Gilf Kebir Plateau, Egypt

age slope of about 35 degrees. They become steeper towards the top, culminating in a vertical, or nearly vertical, lip. (R. F. Peel, *Geographical Journal*, Vol. 93, page 303, 1939.)

The head of each wadi is very abrupt, often ending in a sheer cliff. Examples of these wadis include Hamra (Plate 8.13), Talh Abd el Malik, and El Bakht. The floor of some, like Hamra, is dotted with acacia trees and other shrubs, while many, like Talh, show the remnants of drainage channels.

Other examples of plateaus dissected by wadis are: Adrar, Mauritania (Fig. 8.1) and Plateau du Djado, Niger.

## Inland drainage basins

Deserts have no regular surface drainage and any run-off that does occur tends to flow into shallow

Plate 8.13 Wadi Hamra, Gilf Kebir. The average width is about ¾ km and it is bounded by steep cliffs rising 100 m to the level surface of the plateau

basins, if it is not lost on the way through evaporation or percolation. After heavy storms basins may hold small lakes, but high evaporation soon transforms these into salt flats. In the Sahara temporary lakes and salt flats are called sebkhas (Spanish-American name: playa). Some of the larger basins are due to crustal downwarping (page 1), but smaller ones probably originated as blow-outs. Each sebkha is usually fed by a number of converging stream channels. At the edge of the sebkha and in front of the pediment (Fig. 8.12) is a zone of thick alluvial deposits called the peripediment, which is often partly covered by a salt crust. Basins with uplands on one or more sides usually have a bordering fringe of alluvial cones at the mountain foot. Alluvial cones are similar to fans (page 95), but they are built of coarser material and have steeper slopes of up to 15°. If several cones coalesce into a continuous undulating zone of sand and gravel at the foot of the upland, the feature is

called a bajada. Sometimes the bajada may be so extensive that it almost entirely covers the rock pediment, the outer edge of which slopes beneath the alluvial deposits of the peripediment.

### Chemchane Sebkha, Mauritania

The Chemchane Sebkha is a vast salt flat (Plate 8.14), lying in a depression at the foot of the Grand Dhar, a steep sandstone escarpment that borders the Chinguetti Plateau in the Adrar Massif of Mauritania (Mauritania 1:200,000 IGN, sheets NF-28-XII and NF-29-VII). The sebkha is divided into two sections by a resistant dolerite outcrop which crosses the depression diagonally (Fig. 8.13). On the NW of Chemchane is the Erg Makteir, a region of drifting sand with some barchans.

At the base of the 350 m high Grand Dhar are thick debris deposits backed by a coalescing line of alluvial cones. This zone of deposition slopes gradually down to the edge of the sebkha. On the surface the sebkha is not everywhere the same, being smooth and polished in some parts, and cracked or swollen into mounds elsewhere. Several irregular stream channels lead into the sebkha, but they are mostly blocked by sand and only become active after occasional rainstorms. Chemchane, as it is today, has resulted from the drying out of a large

Fig. 8.12 Features characteristic of an inland drainage basin

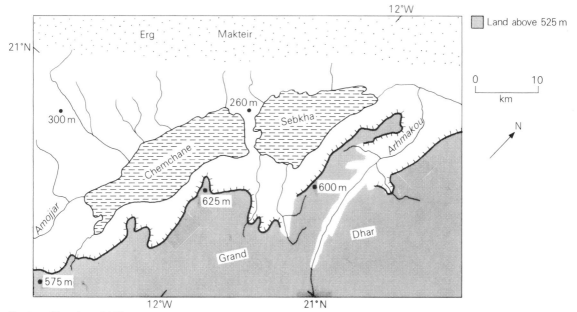

Fig. 8.13 Chemchane Sebkha

Plate 8.14 The dried, cracked surface of the Chemchane Sebkha, Mauritania

lake occupying the floor of a depression, itself eroded by wind deflation over a long period of time, dating back to the Tertiary era.

*Qattara Depression, NW Egypt*

The Qattara Depression (World 1:1,000,000 Map, sheet NH-35) covers more than 15,000 sq km and is about 300 km long from NE to SW. On its north and west the depression is bounded by a steep escarpment, in places over 300 m high (Fig. 8.14). Its southern boundary, however, is indistinct, sloping gradually from the floor 134 m below sea level, up to the surrounding desert plain. The floor of the depression is covered by clays, gravels and salt flats, and is flooded after a rainstorm, but except for the small Lake Maghra there is little permanent surface water.

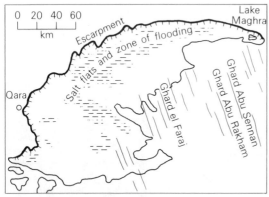

Fig. 8.14 Qattara Depression

The origin of Qattara is entirely erosional and is not related to faulting as is sometimes thought. The rocks in which it has been cut are the weak silts and sands of the Moghra Formation, capped by the resistant Marmarican Limestone. Both rock types are of Miocene age. The depression was initiated by chemical decomposition dissolving heavily jointed and thinner sections of the limestone surface, until the underlying weaker sediments were exposed. Deflation was thus able to wear a hollow in the desert surface, and aided by chemical weathering on the floor and mass movement on the slopes it has worn back the steep Qattara northern wall from the south to its present position. In depth the depression is limited by the water table which forms a deflationary base level.

The many small hollows that lie north of Qattara

have also developed where the limestone thins, while the land between Qattara and the Siwa depression to the west (Fig. 8.4) is underlain by one of the thickest parts of the limestone formation.

South-east of Qattara are long lines of seif dunes, mostly built of sand from the depression and aligned by the prevailing NNW winds. Many of the dunes are named, for example Ghard Abu Sennan. (Ghard means dune.)

Examples of smaller depressions occur at: Siwa, west of Qattara (Fig. 8.4) and Oases of Kharga, Farafra and Dakhla (Fig. 8.4).

## Evidence in the desert of climatic change

In all the African deserts there is strong evidence that in the past the climate was much wetter than it is today. This evidence includes ancient watercourses, old lake features, rock paintings (Plate 1.7), fossil soils, and pollen evidence of trees like pine and cypress. As many as four pluvial periods (page 13) occurred during the Pleistocene, and a number of desert landforms no doubt originated under these more humid climates. Some features, such as the Saharan inselbergs, may even have been formed under earlier savanna climates in the Tertiary.

Radiating in all directions from the Hoggar Mountains and Tassili N'Ajjer (Fig. 8.15) are wide, deep valleys, some over 1000 km long, like the Igharghar, Azaouak, Tilemsi and Tafassasset. The Tilemsi and Azaouak are former tributaries of the Niger. The Tilemsi used to join at Gao and the Azaouak, known in its lower course as the Dallol Bosso, at a point near Niamey. Similar valleys cut other Saharan uplands, for example the Adrar and the Gilf Kebir (Plate 8.13). In view of their tremendous size these wadis have clearly been carved by running water over a succession of pluvial periods, and some probably date back to the Tertiary era, although today they are dry and often filled with sand.

Running east across the Kalahari, from the mountains of Namibia (South West Africa) towards the Makarikari depression, are ancient watercourses called 'mokgachas'. These old valleys, such as the Okwa and Hanehai which begin near Gobabis (Fig.

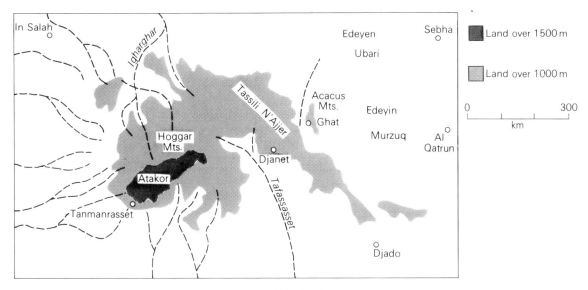

Fig. 8.15 Wadis radiating from the Hoggar Mountains and Tassili N'Ajjer, Sahara

8.2), are now permanently dry. In the south of the Kalahari several major valleys are also dry along their lower reaches, for example the Molopo and the Nossob.

Further evidence of a more humid past climate in the Kalahari is seen in the former lake shoreline, identified by A. T. Grove, that surrounds the Makarikari depression. The centre of the depression is now occupied by the Ntwetwe and Sua pans, but during the Pleistocene the entire area was flooded by a vast lake over 30,000 sq km in size.

In the savanna lands and desert fringes of Africa there is evidence of a time when the climate was much drier than today. This is clear from the ancient sand dunes, now covered with vegetation and cultivation, that occur in certain parts of savanna West Africa far south of the present desert border, notably between Maiduguri and Hadejia in Nigeria. In south-west Sudan fossil dunes, called qoz, exist as far south as 10°N.

### Tassili N'Ajjer

The Tassili N'Ajjer is a huge dissected sandstone plateau lying on the east and NE of the Hoggar Mountains (Fig. 8.15). Tassili is a Saharan word meaning foothills, in this case, of the Hoggar. During the last 40 years the Tassili N'Ajjer has achieved fame from its fantastic collection of rock paintings, which range from 2000 to 7000 years in age. But apart from the paintings, the Tassili is also famous for its remarkable and unique landscape. The plateau is over 600 km long and is made up of a number of separate massifs, such as Jabbaren and Aouanrhet, that are themselves dissected by numerous steep-sided wadis, like the Djerat. The whole has been carved and sculpted into an endless

Plate 8.15 The grotesque rock landscape of the Tassili N'Ajjer, north of Djanet, Algeria

succession of strange and sometimes weird shapes. Tall vertical pillars and pinnacles stand side by side, divided by narrow winding alleys (Plate 8.15). Sheer walls, pierced by rock windows, tower above flat pavements. Here and there natural arches jut out from cliff faces, which in turn are cut by caves and precipitous overhangs. In fact, the scenery of the Tassili has often been likened to some strange city, carved out of monolithic sandstone.

The Tassili landscape is the result of millions of years of differential erosion by wind, water and weathering attacking joints and bedding planes. On all sides the effect of salt crystals and chemical weathering is seen in the rocks that have been pitted and honeycombed into a type of lattice work, while wind erosion is apparent in the grooved and polished surfaces. But by far the greatest work would seem to have been done by running water, under a wetter climate when the now-dry wadis held seasonal streams.

## References

BAGNOLD R. A., 'A Further Journey Through the Libyan Desert', *Geographical Journal*, Vol. 82, pp. 103–129, 1933.

BAGNOLD R. A., PEEL R. F. and others, 'An Expedition to Gilf Kebir and 'Uweinat', *Geographical Journal*, Vol. 93, pp. 295–307, 1939.

CAPOT-REY R., 'Borkou et Ounianga', Mémoire 5, Institut de Recherches Sahariennes, Université d'Alger, 1961.

DAVEAU S., MOUSINHO R. and TOUPET C., 'Les Grandes Dépressions Fermées de l'Adrar Mauritanien, Sebkha de Chemchane et Richât', *Bulletin de l'Institut Fondamental d'Afrique Noire*, Série A, Vol. 29, No. 2, pp. 413–446, 1967.

GROVE A. T. and WARREN A., 'Quaternary Landforms and Climate on the South Side of the Sahara', *Geographical Journal*, Vol. 134, pp. 194–208, 1968.

GROVE A. T., 'Landforms and Climatic Change in the Kalahari and Ngamiland', *Geographical Journal*, Vol. 135, pp. 191–212, 1969.

LOGAN R. F., 'The Central Namib Desert', National Academy of Science, National Research Council, Washington DC, Publication No. 758, 1960.

MAINGUET M., 'Le Borkou, aspects d'un modelé éolien', *Annales de Géographie*, Vol. 77, pp. 296–322, 1968.

PEEL R. F., 'The Landscape in Aridity', *Transactions of the Institute of British Geographers*, No. 38, pp. 1–23, 1966.

SAID R., 'New Light on the Origin of the Qattara Depression', *Bulletin de la Société de Géographie d'Égypte*, Vol. 33, pp. 37–44, 1960.

# CHAPTER NINE

# Glacial and periglacial scenery

## Glacial scenery

Glacial scenery results from the effects of ice as an agent of erosion and deposition. In many parts of the world, including Africa, this scenery dates from periods in geological history when large areas were covered by glaciers and icesheets. These periods are referred to as glacial periods or ice ages. Today, ice sheets are limited to Antarctica and Greenland, and glaciers to the high mountains of the world. Periglacial scenery (page 190) develops under near-glacial conditions beyond the margins of glaciers and ice sheets.

The last major ice age to affect the world occurred in the Pleistocene, when much of North America and Eurasia were overcome by great ice sheets thousands of metres thick. At the same time high mountains, including those in Africa, carried extensive glaciers. The Pleistocene ice age ended about 12000 years ago, having lasted nearly 2 million years. During this time climatic changes caused the ice to advance and retreat several times. The periods when the ice retreated, to roughly the position it is today, are called inter-glacial periods. It is important to note that the fall in temperature required to initiate these glaciers was not especially great. It has been estimated that a decrease of 5°C would be sufficient to cause the East African glaciers to advance to their Pleistocene maximum.

### Past ice ages in Africa

Four or five significant glacial periods occurred in Africa before the Pleistocene. The evidence includes glacial rock outcrops, known as tillite, together with outcrops of older rock on which moving ice cut deep scratches or striations.

At least two, and possibly three, ice ages affected central and southern Africa in the Precambrian. The more important of these was during the late Precambrian when large areas of Zaire, Angola and Zambia were covered by ice sheets. In the Ordovician and Silurian periods much of north Africa was glaciated, including what is now the Sahara. Glacial rocks of this age have been discovered in several regions, notably south-east of Aksum in Ethiopia, near Taoudeni in Mali, and in the Hoggar Mountains of Algeria.

During the Carboniferous and Permian periods a major ice age affected central and south Africa. The ice appears to have spread out from a latitude of Zambia, north to Zaire and south to South Africa. The ice deposited great thicknesses of drift which now occurs as tillite rock in several districts, for example in the Lualaba and adjacent valleys west of Lake Tanganyika in Zaire. In South Africa the Dwyka Tillite rock reaches thicknesses of over 500 m and outcrops range from Natal to Cape Town.

It is impossible to account for these ice ages in a now largely tropical continent unless it be assumed that the continents have drifted to their present positions (page 10). The Permo-Carboniferous ice age, which affected all the present southern continents, including India and Antarctica, has long been a major item of evidence for continental drift. These continents (Fig. 9.1) must then have formed a single land mass (Gondwanaland) sited over the south pole, while at the same time Europe and North America experienced a tropical climate.

The influence of the Pleistocene ice age on Africa was limited to the high mountains. Evidence of Pleistocene glaciers appears in the moraines and

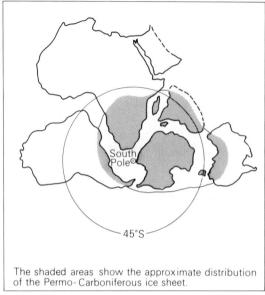

The shaded areas show the approximate distribution of the Permo-Carboniferous ice sheet.

Fig. 9.1 Distribution of the southern continents in late Carboniferous times after the break-up of the former Gondwanaland land mass

cirques in the High Atlas of Morocco and the Semien Mountains of Ethiopia. Moraine ridges also show that ice occupied the Aberdare Range and Mount Elgon, while in South Africa periglacial conditions were dominant in the Drakensberg.

## Ice formation

Glacier ice is an extremely hard substance and originates from the accumulation of snow in areas, such as high mountains, where more snow falls than melts each year. Snow is composed of minute ice crystals separated by air spaces. As it accumulates in hollows on mountains sides the increased weight and pressure of additional snowfalls cause some of the underlying crystals to melt. Water percolates down, filling the air spaces before refreezing occurs. In this way the snow is slowly transformed into a compacted granular mass called firn (névé). As more and more snow falls and the depth of firn increases, the lower part changes into a solid mass of interlocking ice grains from which all air has been squeezed out. This is glacier ice.

## Glaciers

A glacier is a mass of ice largely confined within a valley. Valley glaciers originate from centres of ice accumulation in mountain areas, and under the influence of gravity they move down-slope to occupy pre-existing valleys. Small glaciers that lie in depressions on the mountain side are called cirque glaciers. A very large glacier made up of several valley glaciers coalescing as they reach lower ground is called a piedmont glacier (e.g. Oates Piedmont Glacier, Antarctica).

Whether a glacier advances or retreats depends on the amount of ablation or wastage relative to accumulation. Ablation is chiefly by melting, and involves both melting back at the glacier snout or front and thinning along the length of the glacier. If ablation exceeds accumulation then the glacier will retreat.

## Ice sheets

An ice sheet is a continuous area of ice completely covering the land surface. The Antarctic Ice Sheet is the largest in existence, extending over 13 million sq km, and reaching up to 4000 m in thickness. The Antarctic and Greenland ice sheets today account for over 95% of the world's ice. The mountain peaks that project above the surface of the ice sheets are called nunataks.

On the edge of ice sheets there are usually several outlet glaciers radiating from the main ice mass, for example the Beardmore Glacier in Antarctica, which is over 35 km wide and about 200 km long.

### Glaciers in Africa

Annual snowfalls take place in most of the highlands of Africa, notably the Atlas Mountains (Plate 2.10), the Cape Ranges, the Ethiopian Highlands, the Drakensberg, the Ruwenzori, Mount Kenya and Mount Kilimanjaro, but it is only on the last three that permanent glaciers exist. These present glaciers, however, are but remnants of much bigger features that existed in the Pleistocene. Evidence from glacial deposits on Mounts Kenya and Kilimanjaro shows that the ice advanced several times and glaciers descended to between 2000 and 3000 m.

Plate 9.1 Lac Vert, Ruwenzori, lying in a rock basin to the west of Mt Stanley. On the slopes above are the Alexandra and West Stanley Glaciers

Batian
Nelion
Pt. John
Lewis glacier
Moraine ridge

Plate 9.2 The upper Teleki valley, Mt Kenya. In the centre are the twin peaks of Batian and Nelion, with the Diamond Glacier between. The peak on the right is Point John. The lower glaciers from left to right are the Tyndall, the Darwin and the Lewis

Now they do not come below 4000 m and are still retreating.

Of the three remaining glaciated mountains in Africa, Ruwenzori is by far the most significant, containing 37 glaciers, mainly on Mount's Speke, Stanley and Baker (Plate 9.1). Some of these are valley glaciers, like the Speke and the Moore, while others, like the Northeast Margherita, are cirque glaciers.

In contrast, Mount Kenya (Plate 9.2) has only 10 small cirque glaciers, of which the Lewis is the largest. But even the Lewis is extremely small when compared with such giants as the Beardmore in Antarctica (page 172), for it is less than 1350 m long.

Kilimanjaro has a small ice cap on Kibo (Plate 9.9) with short radiating glaciers, the largest of which is the Penck. Ruwenzori has the greatest number of glaciers largely due to its much higher precipitation. Conversely, the limited extent of Kilimanjaro's ice cap reflects the low annual precipitation.

## Ice movement

The precise mechanism by which ice moves or flows is not entirely understood, but it probably involves a combination of processes, of which the following are the most important.

1 Plastic flowage: ice has plastic properties and under stress, caused by the great weight, it is able to move en masse like a viscous liquid. The pressure within a glacier increases with depth and, as ice grains are pressed together, melting occurs along their boundaries, encouraging intergranular slipping movements.
2 Basal slip, by which the ice slips and slides over the underlying rock.
3 Internal shearing, a movement similar to rock faulting involving differential sliding along planes within the ice body caused by the tremendous downslope pressure.

The speed at which ice moves varies with the gradient and the ice thickness, but it is fastest in valley glaciers. Some glaciers advance several metres a day, while others move less than one metre a year. Movement varies within a glacier, but is greatest in the centre near the surface, where friction is least.

Different rates of movement in the ice cause deep cracks, called crevasses, to form on the glacier surface. Crevasses are common where the bed steepens, as for example on the middle part of the Tyndall Glacier on Mount Kenya and at the snout of the Margherita Glacier on Ruwenzori.

## The work of ice

The effect of ice on landscape is important both directly and indirectly. Directly ice is a major agent of erosion, transportation and deposition. Indirectly ice affects landscape through the meltwaters that issue from the ice front.

All material transported and deposited by ice is called till (moraine). This includes material carried on, within and beneath the ice, ranging from sands, clays and gravels, to rocks and huge boulders. Materials transported and deposited by meltwater from the ice are called fluvioglacial deposits. Glacial meltwater deposits materials in a stratified manner, in contrast to the unstratified deposition by ice.

Till and fluvioglacial deposits are collectively known as glacial drift. In areas formerly covered by large ice sheets the depth of drift may reach hundreds of metres.

Ice erosion involves the 2 processes of plucking and abrasion, of which the former is the most important. Plucking is a quarrying process by which parts of the underlying rock are frozen into the base of the ice and pulled away. It is most effective on well-jointed rocks where meltwater can freeze into the cracks. Particularly important is pressure release jointing (page 6), which appears to be very effective in glacial plucking. Glacial abrasion is a grinding process in which stones and boulders frozen into the moving ice are dragged over the underlying rocks polishing and scratching the surface.

Both ice sheets and glaciers erode the landscape, but the latter produce more dramatic results. If ice sheets are spread over areas of limited relief they may do little more than remove the pre-glacial weathered layer. Glaciers, on the other hand, generally erode more deeply and have the effect of increasing the relief.

Erosion by valley glaciers depends on many factors, such as resistance of the rock, the speed of the glacier, the thickness and weight of the ice, and the availability of rock debris as erosive tools. Erosion increases with the speed and weight of the glacier. The weight of the ice exerts pressures of hundreds of tons per sq metre on the underlying rocks. Erosion is also greater where rocks are weak and there is a large supply of debris.

The till carried by a glacier is derived partly from erosion, and partly from rockfalls onto the ice caused by frost weathering on the slopes above the glacier.

When considering any glacial landform it is important to remember that its formation has taken hundreds of thousands of years, during which time the ice has retreated and advanced several times. Hence its form also owes something to the non-glacial and periglacial (page 190) processes operating during the inter-glacial periods, as well as to the work of ice in the glacial periods.

Evidence for the largescale work of ice, erosive and depositional, can be seen in both mountain and lowland regions. But erosive activity tends to be more apparent in highland areas, while by far the largest collections of depositional forms occur in glaciated lowlands.

## Landforms of glaciated lowlands

### Ice eroded plain

Continental ice sheets cover the entire landscape, and their overall effect is to produce a subdued and rounded topography. The ice scours and scrapes the land surface, carving over-deepened basins in areas of weak rock, and scratching and smoothing resistant outcrops. Pre-glacial valleys may be enlarged and deepened. The former cover of weathered rock is removed, to be redeposited later as glacial drift. During the ice retreat varying amounts of drift are deposited over large areas.

The post-glacial scene therefore includes both depositional and erosional features. The drainage is generally very confused, often with considerable stream diversion due to the breaching of watersheds by ice and the irregular deposition of drift. Rock basins are occupied by lakes, while other lakes form behind drift barriers.

Numerous resistant rock masses, called roches moutonnées, rise above the plain, varying in size from small outcrops to quite large hills. Their sides and upstream end have been smoothed by the ice, but the downstream end is steep and irregular due to plucking (Fig. 9.2). The resistance of many roches moutonnées is due to widely spaced jointing in the rock mass.

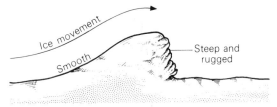

Fig. 9.2 Roche moutonnée

Other resistant rock outcrops obstructed the ice movement and protected weaker rocks on the downstream side from erosion. The result is a rock mass with an elongated 'tail' of weak rocks in the lee. The combined landform is known as a crag and tail (Fig. 9.3). Sometimes the tail is a constructive formation of deposited glacial till. Both crag and tail and roche moutonnée are also to be found in glaciated uplands.

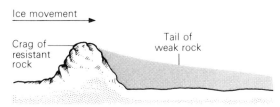

Fig. 9.3 Crag and tail

### Glacial deposition landforms

1 Composed of unstratified till deposited by ice:
   a) **Till plain** – an extensive area of monotonous relief. Former valleys and hills are buried by a thick layer of till deposited beneath the ice sheet and plastered onto the sub-glacial surface by the pressure of the ice. The till content varies according to its place of origin. Many rocks were carried great distances and those that differ from the local rock are called erratics. Most till plains are probably the

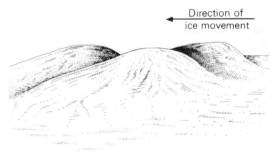

Direction of
ice movement

Fig. 9.4 Drumlins

result of accumulated deposition during several ice advances. Very flat plains usually have a high clay content. Bouldery till produces a hummocky landscape with numerous small hills separated by kettle depressions, sometimes holding lakes. Kettles form from detached ice blocks enclosed within the till after the retreat of the main ice sheet. Subsequent melting of an ice block leaves a depression. Lakes also form in hollows caused by the irregular deposition of drift (page 232).

  b) **Drumlin** – a low, rounded, elongate hill, usually up to 1 km long and about 30 m high. Drumlins occur on till plains in large groups or swarms, aligned to the direction of former ice advance. Most have somewhat steeper upstream ends and consist entirely of till, but some have a rock core (Fig. 9.4). They are formed beneath the ice either by the remoulding of older till, or by the ice shaping till that it has deposited irregularly. They are common in areas where the till plain suddenly widens. Widening would have caused the ice to spread out and form crevasses (page 174), and it may be that crevasses determine the location of drumlins beneath ice sheets.

  c) **Terminal moraine** – an irregular mound or ridge of drift, sometimes extending across country as a belt of low hills for hundreds of kilometres. It is formed by extensive deposition along the edge of an ice sheet, and may be more than 50 m high. Large well-formed moraines develop where the ice front

remained in one position for a long time due to a balance between melting and ice advance. Moraines may be looped in plan, reflecting the lobate edge of the ice, but often they have been dissected by meltwater streams. Many are composed entirely of till, but the outer side of some have a more gently sloping profile built of poorly-sorted fluvio-glacial deposits. A series of roughly parallel moraines mark stages in the ice retreat or readvance, These are recessional moraines.

2 Composed of stratified fluvio-glacial material deposited by meltwater:
  a) **Esker** – a long winding, steep-sided ridge of sand and gravel. It may be more than 30 m high and extend across country for many kilometres, even passing over low hills. Most eskers are ice-contact features formed by streams flowing beneath or within the ice. They develop in areas of stagnant ice where streams can maintain permanent sub-glacial tunnels. Water flowing through the enclosed tunnel is under considerable hydrostatic pressure, making it possible to transport a very large load, much of which may be deposited within the tunnel. When the ice melts the gravelly stream beds remain, with their outer parts collapsed after the removal of the ice-contact. Some eskers form in channels on the ice surface and are later let down onto the ground when the ice melts. Other eskers form where streams drop their load in a mound as they emerge from the ice front. When the ice retreats the mound is extended into an elongated ridge. The width of such eskers varies according to how long the ice front was stationary. Eskers are similar in form to terminal moraine ridges, but they differ in being stratified and lying at right angles to any adjacent moraine ridge (Fig. 9.5).

  b) **Kame** – an irregular mound or hill of sand and gravel. Kames vary greatly in shape and size, and occur in isolation or large groups. They are ice-contact features formed in, on, or at the margin of stagnant ice. Some were deposited as deltas in ice-dammed lakes, others formed from the accumulation of debris in crevasses on the ice surface. Most kames show

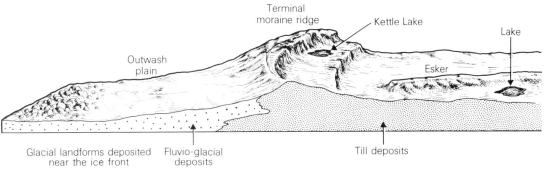

Fig. 9.5 Glacial landforms deposited near the ice front

evidence of the collapse and slumping that occurred when the supporting ice melted.

c) **Kame terrace** – a discontinuous terrace or embankment along a valley side or upland slope. The outer edge may show signs of collapse. Kame terraces are formed by the deposition of sands and gravels in narrow lakes held between the ice front and an adjacent upland. When the ice melts the ice contact edge of the deposit will slump down. Kame terraces may be similar in form to lateral moraine ridges (page 179), but the former differ in being well-stratified.

d) **Outwash plain** – a wide, gently sloping plain of gravels and fine sand. The deposits are often over 50 m thick, and some are pitted by kettles (page 176), which may hold small lakes. Outwash plains are the result of enormous volumes of meltwater spreading materials in great fans beyond the ice front. Braided streams drop coarser gravels first, while fine sands and clays are carried further away.

Few deposition landforms are as simple as outlined here. Most are extremely complex, being the result of both ice and fluvio-glacial deposition that has occurred during several periods of ice advance. Forms deposited by ice may be part-destroyed and remoulded by later fluvio-glacial action.

Finally, it should be noted that meltwater streams are not only depositional agents, but also erosional. Many deep post-glacial valleys were initially carved by powerful floods of water issuing from the front of an ice sheet.

## Landforms of glaciated highlands

Both erosional and depositional features occur in glaciated highlands, but the former are dominant.

a) **Cirque (corrie)** – a steep-sided rock basin, semi-circular in plan, cut into valley heads and mountainsides. Many are very small but some have back walls hundreds of metres high. Some still contain cirque glaciers, others now lie beyond the glaciated zone and are occupied by lakes (page 229). Cirques develop from nivation hollows (page 192). With increasing snowfall a firnfield, and later a cirque glacier, forms in the hollow. Plucking by the glacier, involving meltwater streaming down the walls and freezing into joints, aided by pressure release jointing (page 6), causes the recession of the cirque sides and back wall. The entire process is called basal sapping. The overdeepened basin, with its reverse slope, results from a rotational slipping movement of the glacier with rock debris frozen into its base.

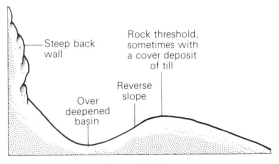

Fig. 9.6 Cross section diagram of a cirque

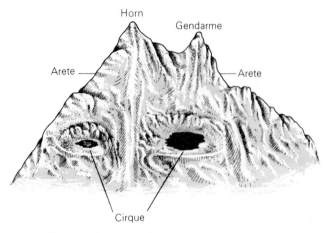

Fig. 9.7 Arêtes, horn and gendarme

b) **Arête** – a narrow, steep-sided rocky ridge, separating two cirques. Formed by the backwall recession of the cirques into the mountain side.

c) **Horn** – a steep-sided, pyramid peak surrounded by cirques, with a radiating system of arêtes. A horn is formed at the junction of arêtes by the backwall recession of two or more cirques on opposing sides of a mountain. Similar to horns are gendarmes, sharp rock pinnacles that rise above arêtes.

d) **Glacial trough** – a broad, flat-bottomed, steep-sided valley, with a roughly U-shaped cross profile. The trough is a complex landform, and within it are found many smaller landforms, both erosional and depositional.

Most glacial troughs are former river valleys that have been modified by ice erosion. It is important to bear this in mind when comparing the ways in which rivers and glaciers carve

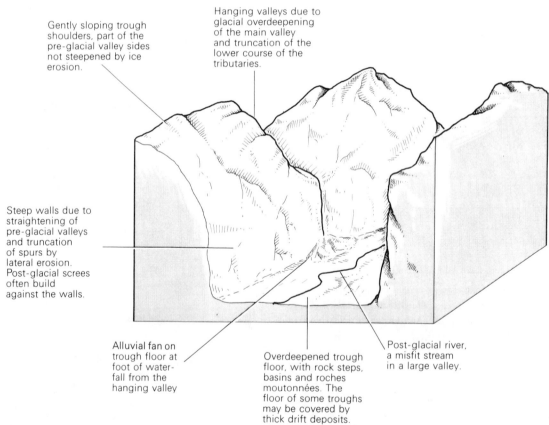

Gently sloping trough shoulders, part of the pre-glacial valley sides not steepened by ice erosion.

Hanging valleys due to glacial overdeepening of the main valley and truncation of the lower course of the tributaries.

Steep walls due to straightening of pre-glacial valleys and truncation of spurs by lateral erosion. Post-glacial screes often build against the walls.

Alluvial fan on trough floor at foot of waterfall from the hanging valley

Overdeepened trough floor, with rock steps, basins and roches moutonnées. The floor of some troughs may be covered by thick drift deposits.

Post-glacial river, a misfit stream in a large valley.

Fig. 9.8 Characteristic features of a glacial trough

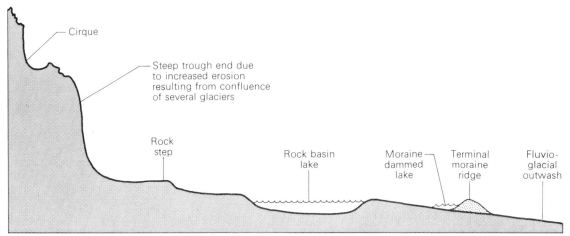

Fig. 9.9 Long profile section of a glacial trough

The long profile may be interrupted by rock steps and rock basins sometimes holding lakes. Rock steps are due to a number of causes, for example, a hard band of rock, or sudden increased erosion due to the confluence of tributary glaciers. Rock basins form where rocks have close-spaced joints and are more easily eroded, or where the valley narrows causing a sudden increase in ice thickness and consequent over-deepening. Many basins are later infilled by drift.

Moraine ridges occur at various places on the trough floor and sides. Crescentic terminal moraine ridges are formed by extensive deposition at the glacier snout where the ice front was stable for a long time. They may soon be dissected by meltwater streams.

Low embankments of till along the sides of the trough floor are lateral moraine ridges. They are formed from deposits collected between the glacier and the valley side. Some lateral moraines extend up the trough sides toward the higher slopes.

Valley floors may also have spreads of fluvio-glacial outwash deposits, particularly in front of a prominent terminal moraine ridge. A long trail of outwash is called a valley train, and it is often soon dissected by post-glacial streams.

Note: Not all these features are found in every glacial trough, and many are not necessarily distinctive of troughs, although they are characteristic.

valleys. The form of troughs may also owe something to inter-glacial river erosion and post-glacial activity. Many characteristics of the trough are a result of the bigger cross-sectional area of a glacier compared with a river. Ice fills nearly the entire valley, and a glaciated valley is, in fact, the equivalent of the river channel, not the river valley. Due to the solid nature of ice, glaciers tend tend to carve much straighter valleys than rivers. Many features of the glacial valley, such as the steep trough end and the

overdeepened main valley with its hanging tributaries, are due to the fact that ice erosion becomes more effective as the glacier moves down-valley and increases its weight from tributary glaciers. In contrast, river erosion is more dependent on the speed of flow. A further difference between ice and river erosion is that ice can be pushed uphill. This is significant in the formation of rock basins.

e) **Fiord** – a deep, narrow arm of the sea with parallel, steep-sided walls. It is a submerged glaciated valley, with many features characteristic of a glacial trough, for example hanging tributary valleys and a U-shaped cross profile. The depth of water in a fiord may exceed 1000 m. The floor often contains many rock basins, while at the seaward end there is usually a rock sill or threshold.

Fiords tend to occur in groups. They are characteristic of western coasts in higher latitudes, and no examples occur in Africa. (In South America, large numbers exist along the coast of Chile, for example Fiordo Aisén and Fiordo Quitralco near Puerto Aisén, while further south, the Messier Channel, which separates Wellington Island from the mainland, has a recorded depth of 1290 m.)

Fiords were formed by glaciers greatly over-deepening former coastal valleys. Their great depth is due to the ability of the glacier to erode

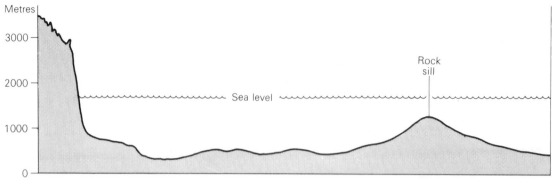

Fig. 9.10 Diagram of a fiord

well below sea level, and is not the result of subsequent drowning, (see page 217). As the glacier moved down-valley its erosive power was gradually reduced and the rock sill marks the limit of effective erosion.

### Glacial troughs on Mount Kenya

Radiating from the central peaks of Mount Kenya are several glacial troughs, for example Teleki, Hobley, Gorges, Mackinder and Hausberg (Kenya 1:25,000 Mount Kenya Special Sheet, 1:50,000 sheet 121/2).

The Gorges Valley lies on the east of the mountain, with its upper end below Simba Col at about 4500 m. Several small streams flow down from the valley headwalls, notably those from Hanging Tarn and Simba Tarn. These streams go to join the Nithi, the main river draining the valley

(Fig. 9.12). The upper part of Gorges Valley, as far down as Vivienne Falls, is divided into sections by 4 rock steps (Fig. 9.11). The highest section, above the first step, is floored by glacio-lacustrine sediments of an old lake (Plate 9.3). At the second step the valley's width narrows sharply, where the Hall Tarns Platform projects southwards. This platform is built of a resistant trachyte rock. Its level surface has been ice-scoured, leaving several small rock basins, some of which hold the Hall Tarns. North of the tarns (lakes) is part of a lateral moraine ridge, about 75 m high. On the second rock step of the Gorges Valley the Nithi River is interrupted by a small waterfall. Below this fall, the valley has been over-deepened to form a rock basin, now occupied by Lake Michaelson (Plate 9.3). North west of this lake the land slopes steeply up to form the cliffs on the south edge of the Hall Tarns Platform.

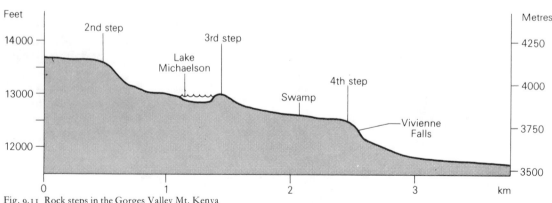

Fig. 9.11 Rock steps in the Gorges Valley Mt. Kenya

Plate 9.3 View east down the upper Gorges valley, Mt Kenya, from the slope below Simba Tarn. The lower part of the valley is obscured by cloud

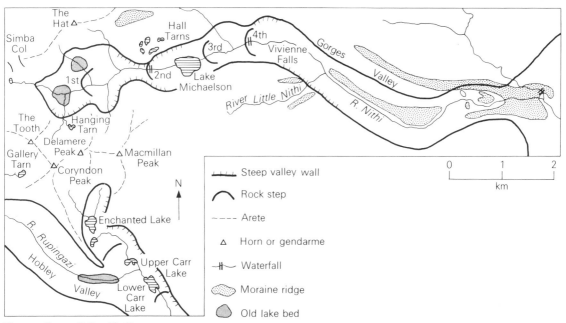

Fig. 9.12 Gorges Valley, Mt. Kenya

The third rock step of Gorges Valley occurs immediately down valley of Lake Michaelson and is less steep than the first two. Below the step the valley floor is flat and partly filled by a swamp. This is followed by the fourth step at Vivienne Falls, beyond which the floor slopes gently for 5½ km as far as Nithi Falls.

Somewhat less than halfway between Vivienne Falls and Nithi Falls, the Nithi River is joined by one of its more important right-bank tributaries, here called the Little Nithi, which flows steeply into the main trough from a hanging valley.

Below the confluence of the Nithi and Little Nithi there are several lateral and terminal moraine ridges. These ridges contain many erratic boulders. Elsewhere on the generally smooth valley floor there are ice-scratched roches moutonnées. At the lower end of the valley the Nithi River cuts through the

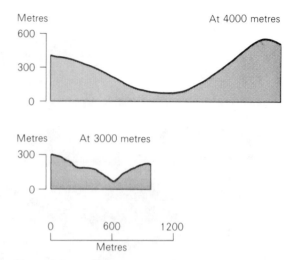

Fig. 9.13 Cross profiles of the Teleki Valley, Mt. Kenya, to show a U-shaped glacial profile at 4000 m in contrast to the narrower V-profile of the river-carved valley at 3000 m

Plate 9.4 The open U-shaped profile of the Teleki valley, Mt Kenya, looking SW from near the snout of the Lewis Glacier

arcuate terminal moraine ridges in a narrow trench, some 25 m deep.

The Teleki Valley lies west of Mount Kenya's central peaks. It too provides an example of a glacial trough. But it differs from the Gorges Valley in several ways, being more open and less steep-sided (Plate 9.4). It contains no significant rock steps, and the slopes above the valley have been smoothed and rounded by ice. All this contrasts with the cirques and serrated rock ridges of the central peaks (Plate 9.5). The Teleki Valley was formerly occupied by the Lewis Glacier, and evidence of its retreat is seen in the occurrence of lateral and terminal moraines, especially toward the upper part (Fig. 9.15). Also the floor of the upper valley is covered by a deposit of fluvio-glacial outwash. The youngest of the terminal moraines occurs at the very head of the valley, as a series of arcuate ridges below the snout of the Lewis Glacier. There are 2 main ridges which are about 3 to 5 metres high, together with a number of smaller ones. On their south side these

terminal moraines join to form one lateral moraine ridge (Plate 9.6). A similar, but smaller, pattern has developed below the Tyndall Glacier, and Tyndall Tarn is a moraine-dammed lake (Fig. 11.2).

### Cirques, arêtes and horns on Mount Kenya

In the area immediately surrounding Mount Kenya's central peaks there are numerous radiating arêtes and over-deepened cirques. The arêtes often end in pinnacle-like gendarmes, notably Point John (Plate 9.2) and Midget Peak at the end of the large arete extending south from Nelion (Fig. 9.15).

A perfect example of a cirque is that occupied by Teleki Tarn (Fig. 9.15). The cirque lies at the upper end of the Teleki Valley (Plate 9.5), from which it is separated by a steep cliff. It is surrounded by sharp-edged arêtes, with four important peaks at junction points: Shipton, Tilman, Grigg and Sommerfelt. The arêtes have developed from the headwall recession of the adjacent valleys and cirques for example Hobley and Hidden Tarn.

Plate 9.5 Mt Kenya from the SW, viewed from near Sommerfelt Peak on the south side of Teleki cirque. Teleki Tarn is not visible, but lies in the dark area below Shipton Peak

Plate 9.6 The central peaks area of Mt Kenya, with Point Melhuish on the left, Batian and Nelion centre background, and the Lewis Glacier right. Bottom right is part of the lateral moraine ridge bordering the glacier

South-west of Teleki cirque is Hohnel cirque, which is situated at the extreme head of the Hohnel Valley. It is similar to Teleki cirque, but is somewhat larger. Lake Hohnel, on the floor of the cirque, is an example of a rock basin lake with its level raised by a large terminal moraine ridge on the rock sill.

### Glacial troughs on Ruwenzori

The glacial troughs on Ruwenzori are in general deeper and more steep-sided than those on Mount Kenya (Uganda 1:50,000 sheet 65/2, 1:25,000 Ruwenzori Special Sheet). Among the most notable are the Mubuku and Bujuku.

The Bujuku has its head between Mount's Speke and Stanley, below the Stuhlman Pass (Fig. 9.14). During the glacial maximum the Speke, Stanley and

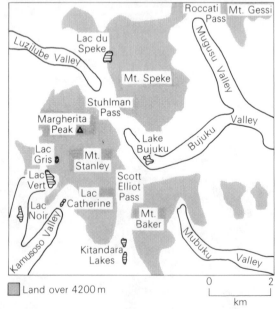

Fig. 9.14 The central area of Ruwenzori showing the main glacial troughs

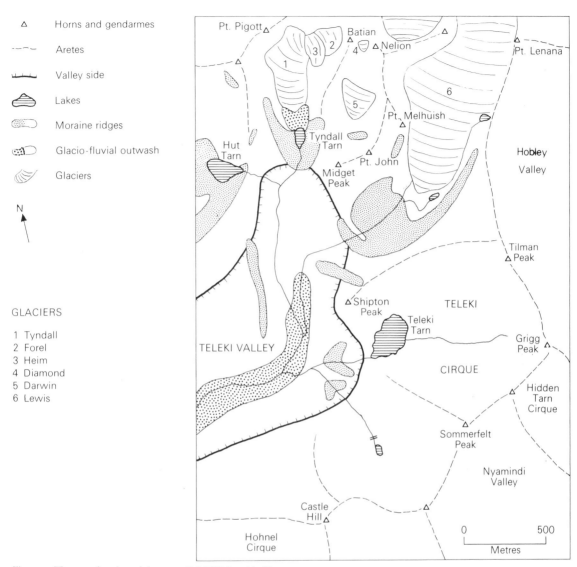

Fig. 9.15 The central peaks and the upper Teleki Valley, Mt. Kenya

Legend:

△ Horns and gendarmes

--- Aretes

Valley side

Lakes

Moraine ridges

Glacio-fluvial outwash

Glaciers

N

GLACIERS

1 Tyndall
2 Forel
3 Heim
4 Diamond
5 Darwin
6 Lewis

Map labels: Pt. Pigott, Batian, Nelion, Pt. Lenana, Pt. Melhuish, Hobley Valley, Hut Tarn, Tyndall Tarn, Pt. John, Midget Peak, Tilman Peak, Shipton Peak, Teleki Tarn, TELEKI, TELEKI VALLEY, CIRQUE, Grigg Peak, Hidden Tarn Cirque, Sommerfelt Peak, Nyamindi Valley, Castle Hill, Hohnel Cirque

0    500
Metres

other glaciers joined up to flow down the Bujuku Valley. The walls of the valley are steep, especially in the upper part, (Plate 9.7) while the floor is occupied by sections of swamp interrupted by rock steps. A particularly marked step occurs above Bigo Hut. At various points along the trough there are small recessional moraine ridges marking the stages of ice retreat.

The Mubuku Valley begins on Mount Baker, and the Mubuku Stream takes its source from the now almost stagnant Moore Glacier. The long profile of the Mubuku Valley is broken by a series of rock steps, the first of which lies some 200 m in front of the glacier snout. Between the glacier and this first step, the valley floor is strewn with an elongated moraine ridge and several ice-scratched rocks, including a large roche moutonnée (Plate 9.8). Along the length of the main valley various lateral and recessional moraine ridges have been identified, and at the lower end, near the junction of the Mubuku

Plate 9.7 Upper Bujuku valley, Ruwenzori, viewed from below the Stuhlman Pass. Mt Baker rises steeply in the background, below which part of Lake Bujuku can just be seen

Plate 9.8 Large roche moutonnée, over 7 m long, on the rock strewn floor of the upper Mubuku valley, Ruwenzori

and Bujuku Rivers, there is a large terminal moraine. This moraine covers a considerable area, extending upstream from the Nyabitaba (Nyinabita) Hut, and Lake Mahoma fills a kettle in the moraine (page 233).

### Terminal moraine ridges on Ruwenzori

Lac Noir (Plate 11.2) and Lac Vert (Plate 9.1) lie in over-deepened rock basins in the upper part of the Kamusoso Valley, west of Mount Stanley (Fig. 9.16). This valley was once occupied by the West Stanley Glacier and in many places along its length there are moraine ridges and glacially striated rocks.

Lac Vert is one of the largest Ruwenzori lakes and is about 25 m deep, but the elongated Lac Noir is less than 10 m deep. Approximately 600 m upvalley from Lac Vert is the much smaller Lac Gris. This almost circular lake lies in a perfectly-formed cirque dammed by a moraine ridge (Fig. 9.16). Between Lac Vert and Lac Gris there are 4 crescent-shaped moraine ridges, the largest, the first, is up to 8 m high. The ridges consist mainly of bouldery till and are covered by groundsels and other plants. Above the steep back wall of the Lac Gris cirque the valley continues for some 500 m. On the floor are

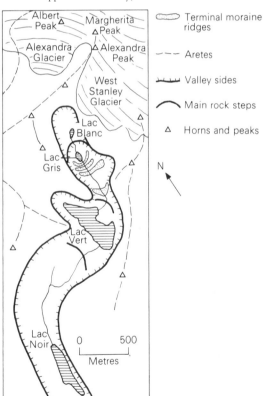

Fig. 9.16 Lakes and moraine ridges in the upper Kamusoso Valley, Ruwenzori

moraine ridges and ice-scratched boulders. Lac Blanc occupies one of several small rock basins in this upper reach of the valley.

*Other examples of glacial features*

*Mount Kenya* – Carr Lakes (Plate 11.1) occupy over-deepened rock basins. Cirques, including those occupied by Hidden Tarn, Oblong Tarn, Hanging Tarn (Plate 9.3), Hausberg Tarn, Gallery Tarn and Emerald Tarn. Horns, occurring at the intersections of arêtes bounding the cirques and valley headwalls between the upper Gorges and Hobley valleys, for example Coryndon Peak and Delamere Peak. In the same area the Tooth is a good example of a gendarme (Fig. 9.12).

*Ruwenzori* – Cirques including that occupied by Lac du Speke (Fig. 11.1) and that lying below Weismann Peak at the head of the Kuruguta Valley on Mount Luigi di Savoia. Two very large lateral moraine ridges extend down the Nyamgasani valley

for several kilometres to below 3000 m. This valley is on the south side of Mount Luigi di Savoia, south of Mount Baker; Lakes Marion and Dominique in the upper Ruamuli (Ruanoli) Valley are elongated moraine-dammed lakes. The Ruamuli Valley extends north from Mount Gessi beyond the Roccati Pass (Fig. 9.15).

*Kilimajaro* – Glacial troughs flank Kibo's south and west slopes, especially the Karanga and South East Valleys (Fig. 3.11). Down the valleys there are elongated lateral moraines; some small cirques have been cut out of the edge of Mawenzi, especially North Corrie in which lies Mawenzi Tarn. On the slopes south of Mawenzi are many roches moutonnées and striated rock outcrops, notably near the track east of Peter's Hut. Extensive outwash deposits cover much of the Saddle between Kibo and Mawenzi, while till deposited in the lee of some of the parasitic cones (Fig. 3.11) has produced examples of crag and tail (Plate 9.9).

Plate 9.9 View west from below Mawenzi across The Saddle toward the glacier-capped Kibo, summit of Kilimanjaro

## Relict features of the carboniferous glaciation in South Africa

During the Permo-Carboniferous glaciation much of south and central Africa was covered by a huge continental ice sheet. In many areas this ice sheet left features typical of an ice-eroded plain, while elsewhere enormous thicknesses of till and outwash gravels were deposited. However, since this glaciation occurred more than 250 million years ago glacial landscapes similar to those of the Pleistocene have not survived.

Nevertheless certain features do remain and they become evident as the rocks that were later deposited on the glacial drift are gradually removed by erosion, and the underlying drift is re-exposed. Over the millions of years since it was first de-posited the till has been hardened into a rock called tillite, the most important of which is the Dwyka Tillite (Fig. 1.6). Near Laingsburg in Cape Province the depth of this tillite is over 600 m. Embedded in the tillite are numerous erratic boulders, particularly of Bushveld Granite, which have helped determine the direction of ice movement (Fig. 9.17). The Bushveld Granite erratics found near Barkly West must have been carried several hundred kilometres by the ice.

In places erosion is removing the tillite to reveal underlying rock basins, roches moutonnées and glacially striated pavements carved in Precambrian lavas and quartzites. Good examples of roches moutonnées in Ventersdorp lavas occur near Hope-town, while near Riverton and Douglas are some fine striated rock pavements (Plate 9.10).

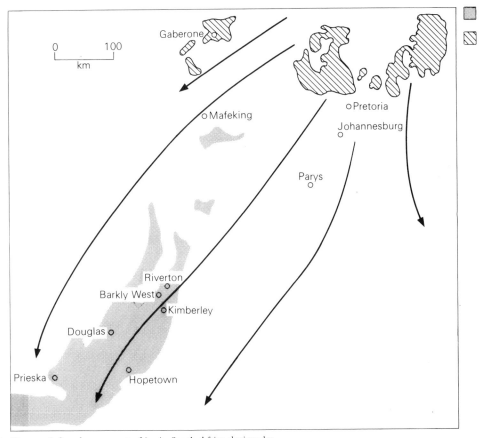

Fig.9.17 Inferred movement of ice in South Africa during the Carboniferous Glaciation. Direction is based on the location of erratics and evidence of glacial striations

Plate 9.10 A glacially polished and striated surface of Precambrian basalt at Nooitgedacht, 22 km NW of Kimberley, South Africa. This glaciated pavement was eroded during the Carboniferous ice age and later covered by deposits of younger sedimentary rocks. These over-lying sediments have since been removed by erosion and the pavement re-exposed

Africa contains no areas of lowland glacial scenery dating from the Pleistocene and the reader is advised to study specific examples in Europe or North America.

## Periglacial scenery

Periglacial landforms develop in high altitude and high latitude areas under a climate where alternate freezing and thawing occurs. In Africa periglacial processes operate on high mountains at altitudes where near-glacial conditions are dominant, notably on Ruwenzori and Mount Kenya. On these equatorial mountains the circumstances vary from the high latitude regions, since the freeze-thaw cycle operates daily rather than seasonally.

The most significant characteristic of the high latitude periglacial areas is that the ground remains permanently frozen except for the upper 1 or 2 metres. This permanently frozen sub-soil and under-lying rock is called the permafrost layer. In parts of the Siberian Tundra it exceeds 500 m in depth. On mountains, however, permafrost is generally non-existent except in the tundra lands. But, whether there is an underlying permafrost zone or not the surface layer plays a special role in periglacial regions. It is subject to daily or seasonal freeze-thaw, and is called the 'active layer'. In the active layer occur the dominant periglacial processes of congelifraction or frost weathering (page 62) and congeliturbation. Congeliturbation is a complex process involving solifluction (pages 70–71) and the development of ground ice. Ground ice varies from small ice needles and lenses (pipkrakes) to large ice wedges, and results from the freezing of water within the active layer. As the water freezes so it expands, heaving up and contorting the ground surface. The extent to which the ground is heaved up is only slight on Africa's high mountains, but in the permafrost lands of the high latitudes it may amount to several metres.

In periglacial areas solifluction (page 70) is a very important process. As the active layer thaws out it becomes a saturated mass capable of sludging down very gentle slopes, especially where there is an underlying impermeable permafrost. The rate of movement varies but up to 10 cm a day has been recorded on some slopes. Deposits of solifluction gravels affect the landscape, since they tend to conceal underlying irregularities in the relief.

### Characteristic features of periglacial areas

### Patterned ground

1 *Stone polygons or nets* – these are polygon-shaped patterns of frost-shattered stones on fairly level ground. They generally contain fine material in the centre, which is sometimes domed up. Polygons vary from the small features of low latitude mountains that are less than 1 m in diameter, to the giant tundra polygons of more than 10 m diameter. But the stone pattern of even the largest polygons rarely penetrates the ground deeper than a metre.

2 *Stone stripes* – these are lines of stones on gentle to moderately steep slopes, separated by finer debris. They are often related to polygons, and there is sometimes an intermediate pattern of elongated polygons or stone garlands (Fig. 9.18). The origin of patterned ground is not fully understood, but is clearly related to differential frost heaving and contraction of the surface layer due to drying and very low temperatures. Over long periods of time these processes lead to a sorting-out of the debris in the surface layer. Ground ice forms beneath stones, which are better heat conductors than finer material, and subsequently heaves the stones to the surface. Polygonal cracks develop in

the frozen ground, because ice contracts the further the temperature falls below 0°C. Ice forms in the cracks, which are gradually enlarged by freeze-thaw. Later, freezing may cause the ground between the cracks to be heaved upwards. This doming tends to cause stones to slip down and concentrate in the polygonal cracks. On slopes steeper than 5° solifluction assists the downslope movement of stones and encourages the development of stone stripes.

### Earth hummocks (thurfur)

These are low mounds, generally less than 1 m high, but often found in large numbers extending over a wide area. Most are covered by tussock grass and sedges. They are thought to form by the congregation of ice lenses beneath the tussocks, followed by gradual frost heaving.

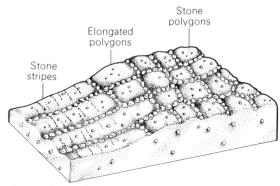

Fig. 9.18 Patterned ground

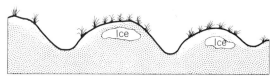

Fig. 9.19 Earth mounds

### Ice mounds (hydro-laccoliths)

These are blister-like mounds varying from about 2 m to 50 m high. In diameter they range from 20 m to 600 m. They are almost entirely limited to the true permafrost areas of the tundra. Up-doming is caused by the growth of a large ice-lens in the active layer. With the onset of winter, freezing occurs from the surface downwards. The lens forms when water is trapped between the newly-frozen surface and the underlying permafrost. Eventually, the ice may melt, causing the overlying mud and gravels to collapse inwards leaving a depression with a raised rim. Small mounds may form in a season, but very large mounds take many years.

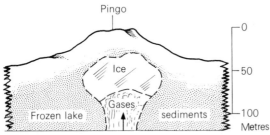

Fig. 9.20 Diagram to show pingo formation

The largest mounds are called pingos, and they are particularly characteristic of the Mackenzie Delta area of northern Canada, where they develop on old lake floors. Groundwater and gases in the lake sediments are trapped by the surrounding permafrost. As the sediments become frozen, so the pressure on the water and gases increases, forcing them to rise toward the surface. This movement, together with the gradual freezing of the water into a large ice body, heave up the underlying frozen and unfrozen ground into a huge dome (Fig. 9.20).

**Nivation hollow**

This is a shallow basin on mountain slopes, often occupied by snow drifts. The enlargement of such hollows by glacial plucking and abrasion leads to the formation of cirques (page 177). Nivation hollows are often less than 100 m across, and tend to form in shaded areas where snow patches persist throughout the year. They originate from the deepening of initial hollows by nivation, the process of weathering beneath and around the edge of a snow patch by frost action and carbonation (see

page 65). The rotted rock is removed by solifluction and meltwater.

*Periglacial features on East African mountains*
Periglacial landforms reach their greatest development in the vast permafrost regions of Siberia and northern Canada. The examples to be observed on Africa's mountains are more limited in extent and smaller in size, but nevertheless they are significant.

The daily freeze-thaw rhythm that operates on Mount Kenya, Mount Kilimanjaro and Ruwenzori promotes the development of needle ice which is the main cause behind the formation of stone polygons and stripes. On Mount Kenya they are most common in the upper valleys between 4350 m and 4800 m. At the head of the Gorges Valley above Lake Michaelson (Fig. 9.12) the polygons are up to 20 cm in diameter. Those on the slopes above the upper Mackinder Valley are smaller and grade into stripes downslope. Stripes are especially well-developed in the upper Teleki Valley below the Lewis Glacier. Each of them is about 10 cm wide. On Kilimanjaro stone stripes are quite common in the area of the Saddle (Fig. 3.11).

Between Lac Vert and Lac Gris on Ruwenzori (Fig. 9.16) there are groups of earth mounds as well as patterned ground. On Mt Kenya a large zone of earth mounds lies east of Lake Ellis between the Mutonga and Hinde Valleys. Some of them are a metre high and $3\frac{1}{2}$ m in diameter.

Elsewhere in Africa fossil periglacial forms have been identified dating from the Pleistocene and earlier periods. In the Drakensberg solifluction slumps occur in several areas, such as above the Sani Pass (Plate 5.12). In the Atlas Mountains and Ethiopian Highlands both active and fossil periglacial forms have been recorded. In the Arak region of Algeria, between Tamanrasset and In Salah (Fig. 8.15), very ancient fossilised pingos, dating from a Silurian glacial age, have recently been discovered.

# References

BAKER B. H., 'Geology of the Mt. Kenya Area', Geological Survey of Kenya Report No. 79, 1967.

DE HEINZELIN J., 'Glacier Recession and Periglacial Phenomena in the Ruwenzori Range', *Journal of Glaciology*, Vol. 2, No. 12, pp. 137–140, 1952.

HASTENRATH S., 'Observations on the Periglacial Morphology of Mts. Kenya and Kilimanjaro, East Africa'. *Zeitschrift für Geomorphologie*, Supp. 16, pp. 161–179, 1973.

DOWNIE C., 'Glaciations of Mt. Kilimanjaro', *Bulletin of the Geological Society of America*, Vol. 75, pp. 1–16, 1964.

DU TOIT A. L., 'The Carboniferous Glaciation of South Africa', *Transactions of the Geological Society of South Africa*, Vol. 24, pp. 188–227, 1922.

WHITTOW J. B., SHEPHERD A., GOLDTHORPE J. E. and TEMPLE P. H., Observations on the Glaciers of the Ruwenzori, *Journal of Glaciology*, Vol. 4, No. 35, pp. 581–616, 1963.

# CHAPTER TEN

# Coastal landforms

The African coastline is notable for its small number of indentations, and compared with its neighbours, Europe and Asia, the continent has a relatively short length of coast, which is remarkable for its generally smooth outline. However, this characteristic of smoothness has not given rise to any lack of major coastal features, which along the 27000 km length demonstrate a wide variety of form. These land-forms range from the high cliffs of the north and south, to the sandbars and lagoons of the west, and the coral reefs of the east. Elsewhere, mangrove swamps alternate with drowned river valleys, while along much of the coast old beaches stand raised high above the present shoreline, evidence of a once higher sea level.

## Factors influencing the shape of the coasts

Variations in the nature of a coastline are essentially the result of a number of important factors that influence all coasts:
1  the work of waves and tidal currents;
2  the height of the land and the nature of the rocks along the coast;
3  whether the coast has been affected by relative movements of sea level;
4  special climatic conditions;
5  the work of man.

### 1. Waves and tidal currents

The main agent of marine erosion, transport and deposition is the breaking action of waves. Waves are undulations in the surface of the sea caused by the blowing of the wind across the water. The friction of the wind on the surface causes individual water particles to rotate in a series of upright circular paths that make a slight forward movement during each orbit (Fig. 10.1). Thus the water itself experiences only a limited displacement, and the chief movement is in the shape of the water, in other words, the waves. Once formed, waves continue their movements long after they have left the area in which they were generated. Two broad groups of waves may be identified.

1  Sea waves – waves actively being initiated by the wind. These are the type in the storm belts of the temperate latitudes. Except for hurricanes, these storm latitudes generate by far the highest waves.
2  Swell – waves that have travelled far beyond their zone of generation. These are the major type in inter-tropical latitudes. Much of the African coast is influenced by swell originally generated by storms in the temperate latitudes, although lo-cally produced waves are also in evidence. Since the dominant mid-latitude storm path is westerly, Africa meets the greatest impact from swell along its western coasts. The south-westerly swell, which is influential from Ivory Coast to South Africa, is particularly significant and is largely generated by storms thousands of kilometres away in the latitudes of the 'roaring forties'.

The nature of the swell has a seasonal vari-ation, reflecting the mid-latitude storm periods. In West Africa, where the swell is south-westerly, waves are highest, about 3 m, in July, while on the Moroccan coast, the north-west swell reaches its maximum in December–January. The eastern coast of Africa is also affected by swell, but here it is less dominant and is partly over-shadowed by waves generated by the onshore trade winds, as well as the occasional hurricane.

## Wave swash and backwash

As waves move towards the coast they enter shallow water. The circular movement of the water particles is then destroyed by friction with the sea-bed. Thus, when the crest of the wave moves forward there is insufficient water in front to support it, hence the wave breaks releasing its energy on the coast. The rush of water up the shore is called the swash, and the retreat of water is the backwash (Plate 10.10).

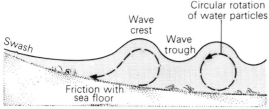

Fig. 10.1 Breaking wave

## Wave refraction

Waves often approach the coast at oblique angles, but as they enter shallower water they are generally refracted by the frictional effect of the sea floor, and their direction is changed such that they break more nearly parallel to the shore. Thus along a coast of alternating headlands and bays, the waves break first at the headlands. Then, as they enter the bays, the wave crests are refracted or bent into a series of curves.

## Longshore drift

Wave refraction, however, never entirely removes the obliqueness, and on coasts where waves break at an angle the swash moves up the shore in the same oblique direction. But the backwash flows back under gravity down the steepest gradient. The effect is that material carried by the waves is moved along the shore in a zig-zag motion. The drift of material along a coast as a result of waves breaking at an angle is called longshore drift (Fig. 10.2).

## Wave erosion

The sea erodes the land mainly by the hydraulic force of the waves hitting the coast, combined with the abrasive action of materials carried by the waves. These materials include sand, pebbles and large rocks. The effectiveness of wave erosion depends on the following:

1 Wave energy – this is related to the size of the waves. Energy is transferred from the wind into the waves and depends especially on (a) the wind speed, (b) the length of time the wind blows and (c) the fetch, or distance over which the wind blows. Storm winds blowing for a long time over a wide area of water generate waves of the greatest energy, and it is in the stormy mid-latitudes that high energy waves are most common. In contrast, the coast of Africa is mainly influenced by long, low, swell waves of medium energy. The most powerful swell is experienced on the west and south coasts, where waves of high to medium energy are recorded. The notable exception to this pattern is the infrequent occurrence of very high energy waves on parts of the east coast generated by occasional hurricanes. The influence of wave energy is most clear during storms when the impact of breaking waves can be as great as 25 tons per sq metre. Such powerful waves hurling rocks and pebbles against the coast cause more change in a few hours than ordinary waves can do in months. In fact, the force of the water itself often helps to loosen rocks for removal, especially when air is compressed into small crevices.

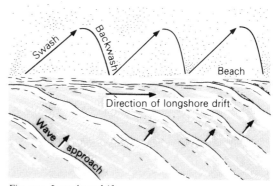

Fig. 10.2 Longshore drift

2 The nature of the rocks along the coast – wave erosion is selective and weak rocks are worn back first, leaving the resistant rocks as headlands. Wave impact is most effective on well-jointed and

unconsolidated rocks, sections of which can be plucked away by the quarrying action of the waves. The height of a coast is important, since a high coastline presents a large amount of material for removal before further erosion is possible, and such coasts may take a long time to erode. On some coasts the rocks are liable to large-scale slumping, to the extent that wave action is almost entirely limited to removal of the debris.

The wearing away, or abrasive work of the waves, depends on a good supply of boulders and coarse sand, which in turn depends on the nature of the coast rocks. In the tropics, the high rate of chemical weathering causes most of the materials to be very fine and thus of limited use as abrasive tools. Limestone rocks are especially susceptible to chemical erosion by solution. On tropical coasts, where coral limestones are often extensive, sea-water solution can play a significant role, especially when abrasion is limited by low wave energy.

3 The exposure of the coast to wave attack – along all coasts wave refraction causes the dominant wave attack to be concentrated on the headlands. But erosion is especially great on exposed coasts, where high energy waves are dominant all the year. Along sheltered coasts, or those with protective off-shore reefs and islands, wave erosion is limited. Coasts with shallow off-shore zones also experience reduced marine attack, since waves are forced to break some distance from the shore.

### Coastal deposition

Breaking waves are important agents of deposition as well as erosion. On some coasts wind is also a depositional agent.

The many depositional landforms found along coasts are only partly built with materials eroded from the coast itself, and by far the major source of sediment is from rivers. In the tropics, where wave energy is less than in the stormy temperate latitudes, rivers are especially important as suppliers of coastal sediments. Along certain coasts, the on-shore movement of materials from the continental shelf is also significant, or has been in the recent geological past. The most common coastal deposits are sand, mud and shingle. Shingle refers to large masses of

small stones rounded by wave action. It is rare in the tropics, where chemical weathering causes the material supplied by rivers to be very fine.

### Tides and tidal currents

The influence of tides on coastal landforms is limited and relates to the range over which waves can operate, that is, between high and low tide. In the tropics the tidal range is generally smaller than in temperate latitudes. On the West African coast it varies from less than 1 m at Lagos to about 5 m in Guinea. In estuaries and narrow channels the ebb and flow of the tide may set up tidal currents, which sometimes have a powerful scouring effect along the sea bed.

## 2. The height of the land and nature of the rocks

The important points here have already been outlined in the section on waves (page 195). In general, resistant rocks tend to form coastal uplands that project into the sea as headlands. Weak rocks, however, usually appear as coast lowlands, and where they alternate with resistant rocks they form bays between the headlands. Uplands are obviously most likely to occur on coasts where mountains reach the sea, for example the Atlas Mountains and the Cape Ranges.

## 3. Relative movements of sea level

Evidence for changes in sea level are found along most coasts (page 12). The importance of such movements cannot be over-emphasised, particularly as they cause the waves to work at a different level.

## 4. Special climatic conditions

Along some coasts special climatic conditions, either past or present, have caused the development of particular features. The most clear-cut examples are coral and mangrove coasts, which are limited to the tropics. Fiords (page 179), on the other hand, occur only on mountainous west coasts in high latitudes, where Pleistocene glaciers carved U-shaped valleys below sea level.

## 5. The work of man

The dredging of estuaries and the building of drainage canals, breakwaters and artificial harbours has often greatly influenced the natural processes of coastal development.

# Coastal landforms

Most coastal landforms result from the interaction of two or more of the above factors, and many sections of coast include a variety of features, each of which demonstrates the influence of several different factors. However, for purposes of description and explanation, the major coastal landforms will be considered individually.

## Cliffed coasts

A cliff is a steep slope or rock face along the sea coast. Cliffs vary greatly in height, profile and plan. Some exceed several hundred metres in height (Plate 10.1), while others are quite low (Plate 10.4). Some are hundreds of kilometres in length, others occur only on narrow headlands separated by sections of low-lying shore. Whereas most cliffs are fronted by a beach (page 200) or wave-cut platform, several plunge directly into the sea.

The majority of cliffs are areas of active marine erosion, but it is only the cliff base that actually experiences wave attack, for along the upper section sub-aerial processes are dominant, especially weathering, mass wasting and downwash. The concentration of wave attack at the cliff base leads to undermining, and on certain coasts a definite notch is cut in the cliff. Notches are especially well-developed on limestone coasts, where chemical erosion is more effective (Plate 10.7). The gradual undermining of the cliff, combined with sub-aerial

Plate 10.1 The 200 m high vertical cliffs at Cape Point, in the Cape Peninsula, South Africa, built of hard Table Mountain Sandstone. The lower headland beyond is the historic Cape of Good Hope

mass movement, causes the upper part to collapse. The fallen debris is then broken down and removed by the waves, so that in this way the cliff gradually retreats.

As the cliff retreats a wave-cut bench may form through the grinding action of materials swept to and fro by the breaking waves. In time the bench is enlarged into a wave-cut platform (Plate 10.2). Most benches and platforms slope seawards and may only

197

Plate 10.2 Wave cut platform in Devonian sandstones extending seawards from the Castle headland at Accra

be visible at low tide, while some may be almost permanently concealed beneath a beach. The most extensive platforms occur on the high wave energy coasts of temperate latitudes.

Coastal rock platforms are not only formed by abrasion. On tropical coasts, periodic wetting of rocks by sea water and spray, alternating with rapid drying in high temperatures, causes gradual chemical disintegration and the formation of a horizontal platform at about high tide level. The processes involved are known as water-layer weathering. Such platforms are most noticeable on low wave energy coasts where they cannot be destroyed by wave erosion.

The shape of cliffs in plan, height and profile depends on several factors. In plan, the cliffline will tend to be straight where the rocks are uniform over a long distance. An indented coast of headlands and bays develops where resistant and weak rocks alternate. But the plan may also be influenced by the exposure of the coast. Cliffs of uniform rock exposed to high energy waves may have joints and other weaknesses eroded into narrow inlets. In weak, unconsolidated rocks, such as volcanic ash, the cliffline is usually a series of short vertical reaches separated by degraded areas where landslides have occurred. Cliff height depends on the nature of the land along the coast. Very high cliffs form where resistant uplands reach the sea (Plate 10.1). Cliff profiles depend especially on the nature of the rock, whether the cliff is being actively eroded at present, and if it is, whether the coast is exposed or sheltered. Since these factors are all variable there is clearly a wide range of cliff profiles (Fig. 10.3). Along tropical coasts the rate of cliff retreat is often slow due to the lower wave energy of these latitudes and the fine nature of the materials available for abrasion. A high proportion of cliffs in the equatorial zone are vegetated and recession is more likely to be due to mass movement than wave action.

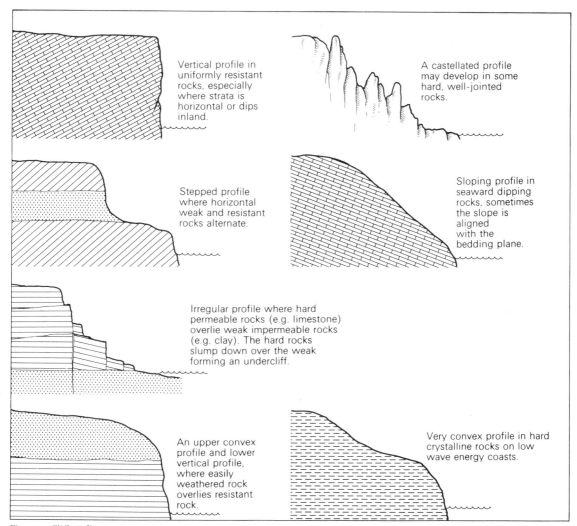

Vertical profile in uniformly resistant rocks, especially where strata is horizontal or dips inland.

A castellated profile may develop in some hard, well-jointed rocks.

Stepped profile where horizontal weak and resistant rocks alternate.

Sloping profile in seaward dipping rocks, sometimes the slope is aligned with the bedding plane.

Irregular profile where hard permeable rocks (e.g. limestone) overlie weak impermeable rocks (e.g. clay). The hard rocks slump down over the weak forming an undercliff.

An upper convex profile and lower vertical profile, where easily weathered rock overlies resistant rock.

Very convex profile in hard crystalline rocks on low wave energy coasts.

Fig. 10.3  Cliff profiles

## Caves, arches and stacks

On certain coasts, as the cliffline is worn back, the waves may carve distinctive features, such as caves, arches and stacks. Caves are holes in the cliff face that develop from waves enlarging an initial weakness in the rock, especially along joints, faults and bedding planes (Plate 10.3). Abrasion and hydraulic force are both vital to cave formation. The impact of breaking waves exerts enormous pressures on rock faces, causing air in crevices to be suddenly compressed. When the water recedes, the air ex-

pands rapidly. This expansion and compression gradually loosens rocks and enlarges cracks. In some caves the force of waves spurting high into the air may weaken joint blocks in the cave roof so much that the roof collapses. The resulting land-form is a vertical shaft connecting the cave with the cliff top and is called a blowhole. A deep, narrow cleft in the cliff, where waves have eroded along a major joint or fault, is known as a geo.

Where a cave develops at the base of a narrow headland projecting from the cliff face, waves may eventually erode right through to form a natural

Plate 10.3 Sea cave near Arniston (Fig. 2.22), South Africa

arch. An isolated rock pillar that fronts a cliffline and has been separated from it by marine erosion is called a stack. It is a remnant of the original cliff. Some stacks are very large and could as well be called islands. Many stacks show clearly the effect of joint control in their formation.

These landforms, caves, arches, stacks, geos and blowholes are not found along all cliff lines, and they tend to be more common on storm wave coasts outside the tropics.

## Beach

A beach is a coastal accumulation of sand or shingle. Beaches are mostly found where the coast is low-lying, but they also occur at the foot of steep cliffs (Plate 4.8). The limits of a beach are determined by low water mark and the maximum level reached by high waves. A beach is constructed and destroyed by the swash and backwash of the waves. Construction is most active during calm weather, when relatively low waves break with a swash that carries sand up the beach. But during storms, waves are higher and break more steeply. This produces a limited swash and a strong backwash that combs sand down the beach.

Beaches vary in size from small ones at the heads of bays, to large features tens of kilometres in length. The width of a beach depends on the supply of sediment available, while the shape of a beach is largely the result of the breaking action of waves which smooth out irregularities. Many beaches are often slightly curved in shape, with a seaward outline that is generally concave, especially in bays, reflecting the influence of wave refraction on entering the bay. The profile of a beach (Fig. 10.4) is constantly being modified, but is approximately

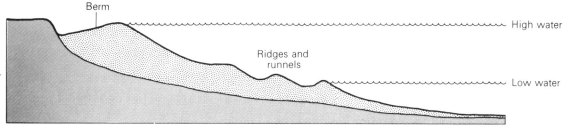

Fig. 10.4 Profile of a beach

concave, with the upper part being steeper than the lower. Shingle beaches are normally higher and steeper than sandy ones, while the steepest sandy beaches tend to occur on swell coasts.

Several minor features also develop on beaches, thereby affecting the overall profile. The more important of these are berms, cusps, ridges and runnels.

### Berm

A bench or ridge with a steep front, found on the upper part of some beaches. Berms are most common on beaches with a steep profile, that is, especially shingle beaches and sandy beaches along swell coasts. Berm formation is influenced by the size of the tidal range, where this is large several berms may develop.

### Cusps

A series of small, horn-shaped projections separated by shallow indentations that face seawards, giving the beach a scalloped appearance (Fig. 10.12). The distance across an identation between the horns of two cusps is rarely greater than 50 m, and the height of a cusp above the adjacent indentation is up to one metre. Cusp formation is most common where waves break parallel to the shore, and very large cusps develop on exposed beaches where large waves are frequent.

### Ridges and runnels (fulls and swales):

Parallel features that develop on the foreshore of a beach and are only visible at low water. They are constructed by breaking waves, and the runnels are often apparent from the shallow pools which occupy them.

### Beach rock

On certain coasts in the tropics and sub-tropics, hard crust-like deposits, called beach rock, project above the sand. Beach rock is composed of sand, shells and pebbles cemented together by calcium carbonate. The exact processes involved in the intensive chemical precipitation needed for its origin are not known, but it is most common in regions with a marked dry season where high rates of evaporation occur. Some outcrops are now being eroded by the sea into a type of rock platform, while elsewhere the encrustations are helping to protect the upper beach from wave attack. In Africa, beach rock is largely confined to the coasts of the Red Sea, Morocco, Kenya and northern Madagascar.

### Barrier beach

A barrier beach is a long sand ridge, parallel to the coast and often separated from it by a lagoon (Fig. 10.5). Barriers that are not actually joined to the coast are called barrier islands. Barrier beaches form from the movement inland of underwater offshore sand bars. These bars are built up from material that is either transported along the shallow offshore zone or is thrown up by waves breaking some distance from the shore. Further deposition increases the height of the bar until it appears above sea level, while wave action causes it to gradually migrate towards the land as a barrier beach. Once in position, barrier beaches develop all the typical features of a beach on their seaward side.

The best conditions for the construction of barriers occur along swell coasts, where long, low waves tend to concentrate materials towards the shore. Also important in barrier formation is the existence of a fairly low tidal range.

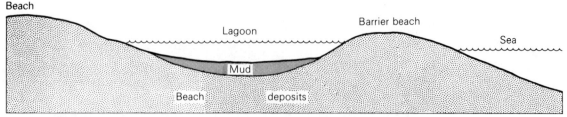

Fig. 10.5 Cross-section of a barrier beach

*Cliffed coasts in Senegal, Algeria and Namibia*

The Cape Verde peninsula of Senegal (Presqu'île du Cap Vert, 1:20,000 sheets Dakar North and South) forms the most westerly point of the African mainland. It separates the relatively straight sand dune coast to the north from the indented Guinea coast to the south. The arm of the peninsula is, in fact, a large tombolo (page 211) chiefly of late Tertiary and Quaternary age, that links the rocky headlands of Cape Manuel and Ngor to the main Senegal coast. The north side of this tombolo is a region of coastal dunes.

The oldest rocks of the peninsula occur at Cape Manuel, where Miocene basalts with strong columnar jointing form cliffs 40 m high (Plate 1.5). Fann Point and the islands of Gorée and the Madeleine are also built of these same basalts. They were originally part of the mainland, but have become separated from Cape Manuel by long-continued erosion. Between Cape Manuel and Fann Point, Eocene sedimentary rocks outcrop. They are chiefly weak calcareous mudstones, and are partly capped by a fossilised lateritic duricrust. The mudstones are more easily eroded than the basalts, and they form the low, pale yellow cliffs that surround Madeleine Bay (Plate 10.4). On the south side of the bay, undermining of the mudstones has caused slumping of the more resistant duricrust layer.

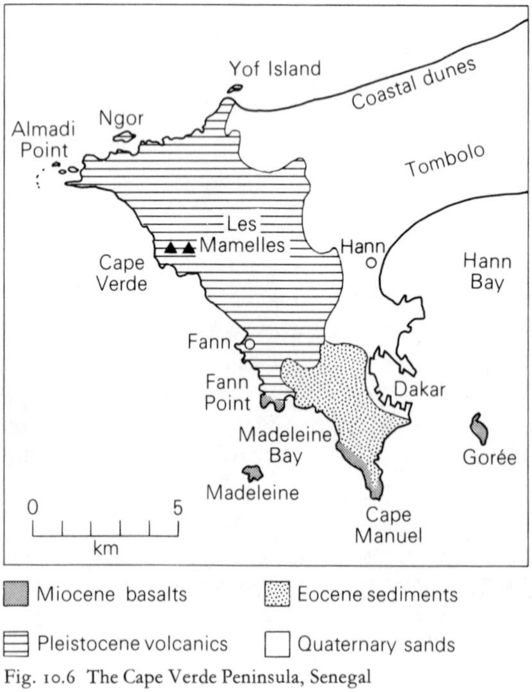

Beyond Fann Point, as far north as Ngor, is an extensive area of basalt and dolerite lavas that resulted from the eruptions of the small Pleistocene volcanoes of Les Mamelles. Marine erosion has gradually cut back into the flanks of the volcanoes, producing the steep cliff face that now forms Cape Verde. South of Cape Verde, towards Fann, the dolerite cliffs become lower, while in the extreme NW the Almadi Reef, a chain of rocks and small stacks, extends seawards from Almadi Point.

At the foot of the dolerite cliffs, especially near Fann and Cape Verde, there is a marked rock bench about 5 m wide (Plate 10.5). This horizontal, pool-covered surface, lies somewhat above high water level and is the result of water layer weathering (page 198) assisted here by the low tidal range, which helps to concentrate the zone of saturation.

Many sections of the Algerian coast demonstrate the influence of different rock types and structure on the shape of the coast. The folded structure of the rocks trends approximately SW–NE (page 29). Since the anticlines of this structure are composed of more resistant rocks they often form headlands,

Fig. 10.6 The Cape Verde Peninsula, Senegal

Miocene basalts   Eocene sediments

Pleistocene volcanics   Quaternary sands

Plate 10.4 Whitish-yellow Eocene sedimentary rocks forming low cliffs at Madeleine Bay, north of Dakar, Senegal

Plate 10.5 Rock bench at the foot of the dolerite cliffs near Fann, Cape Verde Peninsula. Chemical decomposition has greatly enlarged the joints

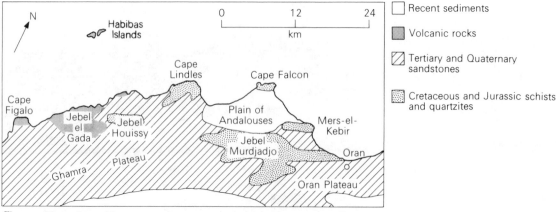

Fig. 10.7 The geology of the coast near Oran, Algeria

while the weaker rocks of the synclines appear as bays. The SW–NE alignment of the folds is reflected in the way the headlands and bays are generally at an angle obliquely to the coast.

Both this structural influence and also the effect of rock type are very apparent near Oran (Ouahran) (Algeria 1 : 50,000 IGN sheets 152 and 153). West of the town, the main capes and headlands, for example Mers-el-Kebir and Cape Falcon (Fig. 10.7), coincide with resistant schists, quartzites and volcanics. Volcanic rocks also form the miniature Habibas Islands. The wide bay of Oran is cut from the weaker Tertiary and Quaternary sandstones on the flank of a large syncline. These same rocks also form the Gulf of Arzeu, north-east of Oran. Behind the rocky Cape Falcon, with its caves and stacks, the land falls away to the low-lying Andalouses Plain which is an area of extensive coastal dunes. The plain is underlain by recent sedimentary rocks and is bordered by long beaches.

Further east along the Algerian coast, between Bejaia and Jijel (Djidjelli) the coast hills slope steeply to the sea, especially near Cape El Aouana (Cape Cavallo). Offshore, there are numerous rocky stacks, such as Grand Cavallo and Petit Cavallo. Among the other famous stacks of the north African coast are the Needles, at Tabarka in Tunisia. These are a group of tall sandstone pillars that stretch out towards a large island in the harbour.

The Namibia coast, south from Luderitz for a distance of about 160 km to Chameis Bay, is rocky, with steep cliffs, stacks and arches (Fig. 8.2). These features largely coincide with the hard Palaeozoic quartzites, dolomites and sandstones of the Nama System. The inlets and bays are mostly eroded in weaker intervening shales. The most famous landmark along this coast is Bogenfels (Plate 10.6), a giant arch of black dolomite, nearly 100 km south of Luderitz. The arch is 70 m high and has been carved by wave erosion removing some inter-bedded softer bands in the dolomite, leaving an upper massive band resting on a supporting pillar.

*Cliffs in Ivory Coast and Madagascar*

The cliffs of western Ivory Coast contrast sharply with the lagoons and sand bars east of Fresco. The cliffs (Ivory Coast 1 : 50,000 IGN sheet NB-29-VI-4 d), are typical of many in the hot, wet tropics, being thickly covered by trees and shrubs, and having a sloping convex profile. Occasionally they rise above 150 m, but for the most part they are below 100 m, and for much of their length they are fronted by a sandy beach. Except on the headlands, such as at Tabou, the cliffs are only being actively eroded by the sea in one short section. This is a zone of horizontally bedded clays and sandstones some 3 km long, west of Fresco. The cliffs here are almost vertical and largely devoid of vegetation. They have a slope of about 50°, and except for the lowest few metres above the beach, are almost entirely covered by vegetation, which on the upper cliff is often thick forest.

Along this coast cliff recession is very slow in both crystalline and sedimentary rocks, and is

Plate 10.6 Bogenfels, the giant natural arch of Palaeozoic dolomite on the Namibia coast, south of Luderitz

chiefly the result of weathering and mass movement, rather than marine erosion. The rocks have been deeply weathered by chemical decomposition, and are very susceptible to slumping. Landslide scars are usually visible on the cliff face, although they soon tend to be recolonised by vegetation. Large boulders from the slumped material sometimes remain on the beach for a considerable time, since the waves seem to be rarely capable of removing them or eroding them. The dense vegetation of this coast is made possible by the higher rainfall totals than the Ghana coast to the east, so reducing the effect of salt from the sea spray.

North of Soalara (Fig. 6.20) in SW Madagascar, cliffs of horizontally bedded Tertiary limestones, nearly 100 m high, plunge straight into the sea. At the cliff base there is an almost continuous notch up to 5 m deep and about 4 m high, with a very marked overhang, called a visor. Wave energy is low here

and the notch would thus seem to be largely a result of concentrated seawater solution. Above the notch the cliff is pitted by the effect of spray from the waves. The upper cliff slopes inland and is covered by a sparse scrub vegetation (Plate 10.7).

### Coast erosion in Ghana and Mauritius

Figs. 10.8(a) and (b) are of the same section of coast, but with a seven-year interval between them. In 1960 the Nkontompo headland, midway between Sekondi and Takoradi in Ghana, was still joined to the cliff by a narrow neck of land (Ghana, Sekondi Town Plan 1:1250 sheet 7). Between 1960 and 1967 the cliffline of Palaeozoic sandstones and shales was eroded back by 20 to 30 metres in the area SE of the headland. Over the same period the waves also cut through the narrow neck and formed the headland into a large stack (Plate 10.8). The main process of cliff recession began in 1963, and by 1965 when the

Plate 10.7  Limestone cliffs, with a deep notch and overhang, near Soalara, SW Madagascar

Plate 10.8  Nkontompo stack, between Takoradi and Sekondi, Ghana

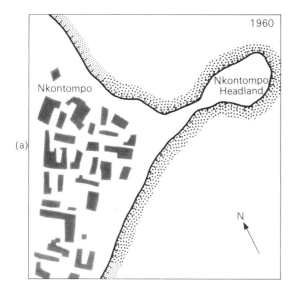

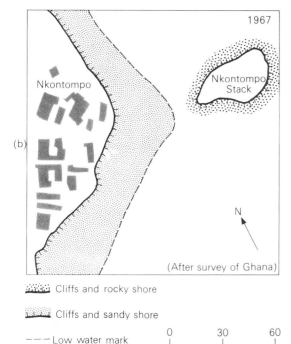

Cliffs and rocky shore

Cliffs and sandy shore

--- Low water mark

```
0        30        60
|_____|_____|
        Metres
```

Fig. 10.8 Maps to show coast erosion near Sekondi, Ghana

the west of the southern Indian Ocean, where approximately 15 hurricanes occur every year. On 25 February 1960 Hurricane Carol passed directly across the island. During the storm wind speeds reached over 225 km per hour, causing very large destructive waves. Further, along those parts of the coast where winds were onshore the level of the sea was temporarily raised by more than half a metre, so increasing the zone of attack. Waves undermined the basalt cliffs in many parts, causing recessions of up to 6 m, while along much of the coast beaches were severely eroded (Fig. 10.9).

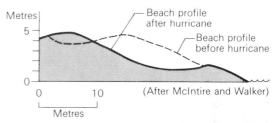

Fig. 10.9 Beach profile at Pointe Lafayette, NE Mauritius (25 km due east of Port Louis), showing the effect of the 1960 hurricane

### Le Souffleur, a blowhole in Mauritius

On the south coast of Mauritius, about $3\frac{1}{2}$ km SE of Savannah, between Mahebourg and Souillac is a remarkable natural blowhole, called Le Souffleur, 'the Blower'. The basalt rocks of this coast have here been eroded into a low headland, joined to the mainland by a narrow neck. The beating of the waves has undermined the headland and formed caves on all sides. In two places, the sea has carved cylindrical openings up through the rock to the surface. The noise of the waves forcing their way up the blowholes can be heard up to 2 km away.

### Lagoons and sandbars of the Togo–Benin coast

A major feature of the West African coast is the series of lagoons and sandy barrier beaches, especially between the deltas of the Volta and the Niger. In Benin (Dahomey), the lagoons are particularly large, for example Nokoué and Ahémé (Togo–Dahomey 1:500,000 IGN sheet Porto Novo). Altogether, lagoons cover more than 20,000 hectares along the Benin coast (Fig. 10.10). To the east, in Nigeria, the name Lagos is derived from the characteristic lagoons of this coast. West of the

stack was created more than ten houses in Nkontompo village had been destroyed.

The erosion in Ghana was spread over several years, but in 1960 Mauritius experienced rapid coast erosion in only a matter of hours. Mauritius lies on

Plate 10.9 The narrow strip of the Sakumo Lagoon west of Accra, near the village of Botianor. The 3 m high barrier beach is clearly visible in the foreground

Volta, the Ghana lagoons are much smaller (Plate 10.9).

The lagoons have been formed by the building up of barrier beaches across drowned river mouths. A rising sea level following the Pleistocene glaciation was the cause of most of the coastal rivers being submerged near their mouths. Some lagoons are transverse to the coast, and their shape reflects the outline of the drowned lower course of a river. Other lagoons, however, are elongated and parallel to the coast. The heights of the sand barriers vary, but most are about 5 m to 10 m above sea level. Some are large, relatively mature features that have been stabilised by roads and permanent settlements.

Most of the material that formed these barriers has been moved along the West African coast by the dominant west–east longshore drift. The estimated drift of sand and other materials each year is about $1\frac{1}{2}$ million tons. Visible proof of this drift can be seen in the accumulated sand against the west side of breakwaters at Takoradi and Cotonou. However,

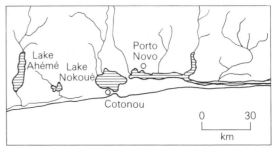

Fig. 10.10 Lagoons of the Benin coast

there are variations to this west–east movement, and in some areas the drift is westwards. This is apparent at the Pra River mouth, east of Takoradi, where a spit (page 210) has been built by material drifting west.

The dominant agent in the building of the barrier beaches is the powerful SW swell, the waves of which are about $1\frac{1}{2}$ m high on the Benin–Nigeria coast.

The present barriers of the Benin coast are

Plate 10.10 Victoria Beach, Lagos, from the landward end of the East Breakwater. Backwash is seen flowing back down the beach to slip under the freshly breaking wave

only the latest in a series that have formed since the Pleistocene glaciation. The old barriers now lie inland, in some areas up to 30 km. They appear as a number of ridges, separated by depressions once occupied by the former lagoons. The rich vegetation in the old lagoon beds contrasts sharply with the sparse cover of the old beach ridges.

Other examples are: Ivory Coast, especially Aby Lagoon and Ebrié Lagoon; Congo and Gabon coasts, for example Ndogo Lagoon; the east coast of Madagascar, south of Tamatave, for example Nosive Lagoon and Ampitabe Lagoon.

### Beach erosion on the Nigerian coast, near Lagos

East of Lagos, on the coast of Victoria Island, marine erosion is causing the large-scale retreat of the shore at Victoria Beach (Plate 10.10) (Nigeria 1:50,000 sheet 279 SE). The erosion began following the construction of breakwaters at Lagos Harbour entrance, built between 1907 and 1916. These breakwaters halt the eastward drift of sand along the coast, which formerly passed across the harbour mouth. The long-term effect of this has been that to the west of the harbour entrance sand has accumulated to the extent that Lighthouse Beach has advanced by over 400 m. East of the entrance, however, measurements show that erosion near the East Breakwater is so great that Victoria Beach has retreated more than 1250 m between 1912 and 1970 (Fig. 10.11).

Recent research by E. J. Usoroh has shown that the erosion is greatest between March and September each year, when storm waves are at a maximum due to the influence of the onshore SW winds. In addition soundings in the off-shore zone show a greater depth near the East Breakwater than in other parts. This increase in depth is thought to be due to the action of eddies caused by the alignment of the breakwaters, and its result has been to steepen the off-shore profile and so intensify wave energy at Victoria Beach.

Various attempts have been made to counteract

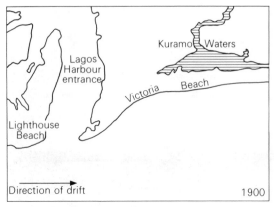
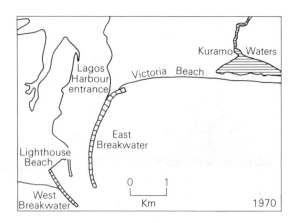

Fig. 10.11 Erosion at Victoria Beach, Lagos

this erosion, of which the most significant have been the inland extension of the East Breakwater at the western end of Victoria Beach, and the regular pumping of more than half a million tons of sand from Lagos Harbour onto the beach each year. Nevertheless, the erosion continues, and the average annual recession between 1967 and 1969 was more than 18 m.

An almost identical example of coast erosion, resulting from harbour installations interfering with the longshore movement of material, can be cited from Durban, on the SE coast of South Africa.

*Beach cusps at Cape Coast, Ghana*
Beach cusps at Cape Coast are well-formed west of the Castle Headland on a beach that stretches in a straight line for 10 km almost as far as Elmina. The cusps are not always continuous, the main interruption being at the barrier across the entrance to the Fosu Lagoon (Fig. 10.12). Annual draining of

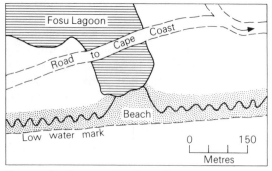

Fig. 10.12 Beach cusps at Cape Coast, Ghana

the lagoon tends to destroy what development does take place at this point. Along most of the beach the cusps are fairly large with the average measurement between their apexes of 30 m (Plate 10.11). Although the cusps are constantly changing in shape, they are nevertheless a relatively permanent feature of this beach.

Examples of runnels and ridges, and berms, can also be identified along many African beaches. A large well-developed berm occurs on the long beach south of Cape Lopez, near Port Gentil, Gabon.

Runnels and ridges are significant features on the foreshore of the wide sandy beach at Sidi-Ferruch, west of Algiers.

## Spit

A spit is an embankment of sand or shingle, attached to the land at one end and extending seaward. Some spits form parallel to the coast and may be tens of kilometres long, while others grow at an angle, often projecting across an estuary. The mouths of rivers and lagoons are sometimes deferred several kilometres by spit growth. Spits develop from the movement of material by longshore drift (page 242), and commonly form where the orientation of the coast changes. On their landward side they are often strengthened by the growth of vegetation. Spit length is generally limited by a reduced supply of material, increasing depth of water, or, in an estuary, the existence of strong tidal currents.

Plate 10.11 Beach cusps at Cape Coast, Ghana. The embayment of a cusp is seen in the centre, with the cusp apex below the prow of the boat. Cape Coast castle headland is visible in the distance

Certain spits have their seaward end curved into a hook shape. This recurvature is common where waves approach from several directions, or where water is deep enough off the spit to allow wave refraction to curve the end.

## Tombolo

A tombolo is a spit that grows out from the coast and links an island to the mainland.

## Cuspate foreland

A cuspate foreland is a large triangular-shaped deposit of sand or shingle projecting seawards. It is formed only rarely, and then largely as a result of the coalescence of two spits growing towards each other at an angle. The foreland is gradually enlarged, by the addition of more material as beach ridges on the outer edge of the spits, while the lagoon enclosed by the spits is slowly silted up.

*Spits in Senegal and Tunisia*

The Langue de Barbarie is an enormous linear sand spit parallel to the coast at the mouth of the Senegal River (Fig. 10.13). It has deflected the river southwards and is more than 40 km long but only about $\frac{1}{2}$ km wide (Plate 10.12). The spit is growing at about 100 m a year, but this has been greater in the past. The main factor in this growth is the regular and powerful NW swell, which is made the more effective by onshore winds. Alignment of the spit is not related to the offshore Canary Current.

In 1658 the spit end was less than 5 km south of St Louis, yet by 1825 it was nearly at the latitude of Gandiole. However, the spit has not always advanced. Between 1884 and 1887 it retreated nearly 4 km. Erosion of the spit end is due to the occurrence in some years of exceptionally strong waves from the SW. These same waves have also been responsible for occasionally cutting gaps in the spit. But the regular southerly drift of material soon fills the gaps (Senegal 1:200,000 IGN sheets NE-28-II, ND-28-XX).

One of the most amazing examples of hooked

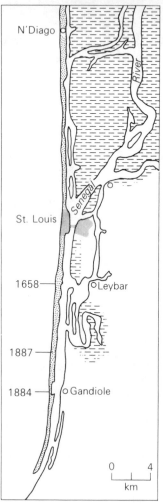

≡≡≡ Mangrove

░░ Sand spit

Dates signify the position of the spit end in that year

N'Diago

St. Louis

1658 — Leybar

1887

1884 — Gandiole

0    4
km

Fig. 10.13 The Langue de Barbarie, Senegal

spit formation was the structure built at the mouth of the Medjerda River in the Gulf of Tunis (Figs. 10.14 and 5.15). Here, material deposited by the river at the delta front between the lagoons of Sebket el Ouafi and Sebket el Bounta was recurved by wave and current action to produce a crescent-shaped formation. However, since 1953 and the construction of an artificial diversion canal, the deposition of sediment by the river has been reduced, with the result that the waves have now almost entirely destroyed the spit.

Other examples of spits are: Walvis Bay spit, Namibia which is nearly 15 km long (Plate 8.8), and is one of several on this coast constructed by north-moving longshore drift; Lobito spit, Angola, which increases its length by up to 20 m a year as a result of the annual northward drift of more than 250,000 cu metres of sediments; Ras Luale, a spit 50 km north of Dar es Salaam on the Tanzanian coast; and Banjul spit, Gambia, where the capital lies on a recurved spit at the extreme east end of the low-lying mangrove-covered sand bank known at St Mary's Island (Plate 10.13).

### Tombolos in Somalia and Sierra Leone

The tombolo at Ras Hafun in Somalia forms the most easterly point of the African mainland (Fig. 10.15). Here a giant sandbar, more than 25 km long, links the coast with a former offshore island.

Plate 10.13 View from the north of St Mary's Island, in the Gambia estuary. Banjul lies on a hooked sandspit at the far end. Mangroves are visible centre right

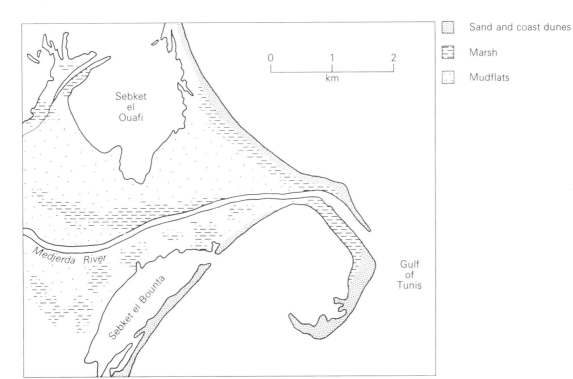

Sand and coast dunes

Marsh

Mudflats

0   1   2
km

Sebket
el
Ouafi

Medjerda River

Sebket el Bounta

Gulf
of
Tunis

Fig. 10.14 Recurved spit of the Medjerda Delta, Tunisia

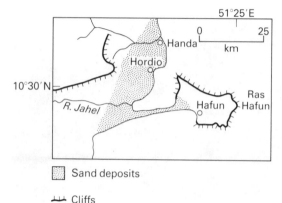

Sand deposits

⊥⊥ Cliffs

Fig. 10.15 Ras Hafun Tombolo, Somalia

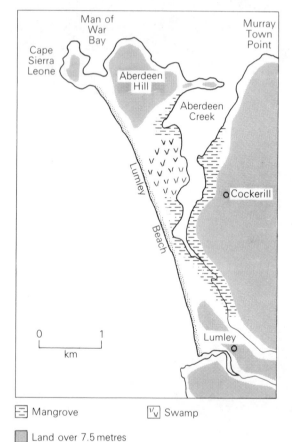

⊟ Mangrove     ᵛᵥ Swamp

▓ Land over 7.5 metres

Fig. 10.16 Lumley Beach Tombolo, Sierra Leone

A similar, but smaller, feature is that which joins the main part of the Sierra Leone Peninsula to the former islands of Aberdeen Hill and Cape Sierra

Leone. The west face of this tombolo, which extends NNW from near Lumley, now forms Lumley Beach, while to the east are areas of swamp and mangrove bordering Aberdeen Creek (Fig. 10.16).

### The cuspate foreland of Point à Larrée, Madagascar

Point à Larrée is a huge triangular-shaped formation of more than 60 barrier beach ridges. It lies in the lee of the elongated island of Sainte Marie on the east coast of Madagascar, about 100 km north of Foulpointe (Madagascar 1:50,000 sheets W.41, W.42, X.41/42). Due to the position of the island, a unique set of wave movements are set up in the narrow channel between it and the mainland. Waves approaching the coast from the east are deflected around each end of Sainte Marie island. The effect is to cause some waves to come in roughly from the SE and some from the NE. The opposing wave directions have gradually built up two series of barrier beaches, that have been extended seawards to form a pointed cuspate foreland (Fig. 10.17). The parallel beaches now appear as rows of successive ridges, and all, except the most recent, are covered by a low scrub vegetation. A narrow linear lagoon follows the axis of the foreland. The north side of Point à Larrée shows the parallel alignment of the

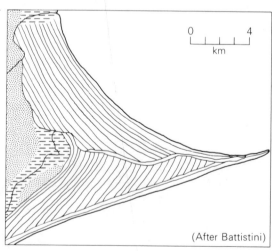

▒ Old beach deposits     ⊟ Marsh

▨ Barrier beach ridges

Fig. 10.17 Point à Larrée, Madagascar

(After Battistini)

beaches with the present coast. But along the south side, past erosion has cut back the extremities of the old ridges, and only the most recent beaches are aligned to the waves.

## Coastal dunes

Sand dunes develop on coasts where winds are predominantly onshore, and are sufficiently strong to move a large supply of sand inland from a wide beach area. They are especially common along arid and semi-arid coasts. The dunes are more limited in area than those of the desert, but they form under similar conditions of wind flow. A factor that is often important in coastal dune formation is the presence of grass and other vegetation to trap and stabilise the sand.

Various types of dune develop along coasts, notably barchans (page 159), transverse dunes (page 160), small dome-like mounds called nebkas, and parabolic dunes. These later form by the central part of the dune migrating inland while the sides are fixed by vegetation, giving a crescent shape with horns pointing to windward.

### Coastal dunes in southern Madagascar

Along the south coast of Madagascar are extensive areas of coastal dunes, some of which are ancient and completely fixed by grass and creeping vine vegetation, while others are still mobile. One of these areas occurs about 50 km west of Fort Dauphin, near the mouth of the Mandrare River (Madagascar 1:100,000 sheet L.62, Plate 10.14). Along the coast itself is a long straight barrier beach, interrupted by a slight gap at the Mandrare mouth (Fig. 10.18). Behind the beach, to the east of the river is the large lagoon of Lake Anony. The active dunes are best developed in the area between the river and the lake, the most widespread type being transverse dunes in two large fields. Behind the barrier beach are many small nebkas less than 5 m in height. All the dunes seem to have been built up by dominant SE winds.

Other examples are: the Atlantic coast of Morocco, for example south of Essaouira (barchans and nebkas) and west of Tiznit (parabolic and transverse); the north shore of the Cape Verde peninsula, Senegal, where SW-trending mobile dunes enclose small fresh water lakes called niayes; and the Red Sea coast of Sudan, near Suakin, where barchans are common.

Plate 10.14 Transverse coastal sand dunes at the mouth of the Mandrare river, south Madagascar. The view is from above Antanandava looking SW towards the barrier beach

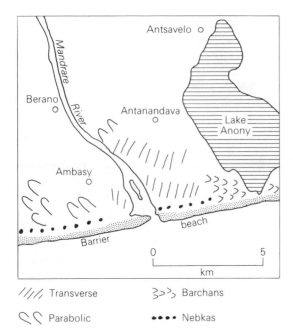

**Symbols legend:**

`////` Transverse  `>>>>` Barchans

`ςς` Parabolic  `••••` Nebkas

These symbols do not represent individual dunes

Fig. 10.18 Coastal dunes near the Mandrare River mouth, Madagascar

## Mangrove swamps and other coast vegetation

Vegetation often plays a major role in coastal deposition by stabilising estuarine mudbanks, barrier beaches, dunes and spits, while the plants themselves often help trap more deposits. On the upper part of beaches and on coastal dunes (page 215) various succulents, herbs and coarse grasses are common. But in areas subject to tidal change, such as coastal creeks and lagoons, the plants are halophytic. Halophytes are plants able to grow near salt water. In the temperate and sub-tropical areas grasses and sedges colonise tidal mudbanks and develop salt marsh, especially in sheltered creeks and estuaries protected from direct wave action.

In equatorial Africa mangrove is the dominant coastal vegetation. Along these hot, wet coasts salt marsh is largely excluded. There are about 30 different mangrove species, all of which are adapted to root themselves in unstable mud and tolerate tidal inundation. Ideally, the annual temperature range should not exceed 5 °C and the average for the coldest month should not fall below 20°C. A dense network of mangrove roots helps trap more mud, and so the mudbanks extend seawards and increase in height.

### Mangrove swamps of the Niger Delta

On the west coast of Africa mangrove occurs in a number of areas ranging from Senegal (Fig. 10.13) to as far south as northern Angola. But by far the largest swamps are in the Niger Delta, where mangrove covers more than 9000 sq km, and extends inland for up to 30 km (Fig. 5.13). Here the swamps reach full development on the fine-grained, soft organic muds deposited by the river. They lie between low and high tide, and form an almost impenetrable zone of trees, criss-crossed by an intricate system of creeks and channels. Annual rainfall in the delta is heavy, exceeding 3500 mm. This is an important factor, causing the mangroves to reach 12 m in height, with many more than twice as high. Some (Avicennia) develop horizontal roots with side shoots growing up through the mud, while others (Rhizophora) have stilt-like roots to support the trunk above water level. The seaward margin of the swamps is fronted by a sandy beach zone. This has an average width of 5 km and consists of a series of parallel ridges that rise to some 4 m above mean sea level.

On the East African and Madagascar coasts the mangroves are visibly similar, but the species are more akin to those in South East Asia than to the West African varieties. Also they extend further south, reaching Maputo in Mozambique.

## Coastal scenery resulting from changes in sea level

Relative movements of sea level can be grouped into negative and positive (page 12). The movements are due to actual changes in sea level itself or to movements of the coast. Sea level changes, eustatic changes, can result from a change in the size of ocean basins caused by large-scale earth movements, or more commonly from a major glaciation abstracting and releasing large quantities of ocean water. Coastal movements involving uplift or de-

pression of the land result from local earth movements or from isostatic adjustments due to unloading or loading of the crust (page 8).

The evidence for relative sea level change is
apparent along all the African coast, some sections
showing a rise and some a fall. The most widespread
cause was eustatic change during the long Pleistocene glaciation. This involved both positive and
negative movements due to the occurrence of
interglacial periods, when the sea level rose. In some
areas local earth movements and isostatic adjustments were superimposed on these changes. The
overall effect is that much of the coast shows
multiple evidence for both negative and positive
movements. Negative movements produce emerged coasts, while positive movements cause submerged coasts.

## Submerged coasts

These are deeply indented with the lower parts of
river valleys forming enlarged estuaries. The
funnel-shaped drowned river valleys are called rias,
which are wider and deeper at their entrances than
inland. The actual shape of a ria coastline will
depend on the height of the land before submergence. Along a submerged upland coast the
former divides between the valleys now form
headlands, with isolated hills appearing as offshore
islands. In time, cliffs develop along the headlands.
On a submerged lowland coast the higher base level
leads to increased deposition by rivers and this helps
in the formation of mudflats and swamps, especially
along ria sides. Submergence of an area with
elongated hills and valleys parallel to the coast
produces a pattern of long narrow inlets and islands
in a longitudinal layout.

Comparison is often made between ria coasts and
fiord coasts (page 179) because of the superficial
similarities. However, fiords are generally very
much larger and deeper than rias, their sides are
often vertical, and they have relatively shallow
entrances. Further, fiord coasts are not directly due
to drowning, while ria coasts are. Fiords were
carved by glaciers able to erode below sea level, so
that even if sea level had not risen they would still be
submerged. Thus, although the depth of rias is

entirely due to coastal submergence, the depth of
fiords is largely a result of glacial overdeepening.

## Emerged coasts

Where a relative fall in sea level leads to the
emergence of a large part of the continental shelf,
the new coast takes the form of a wide, gently
sloping plain. The shoreline is usually fairly even,
with no bays or headlands. On such coasts, the
offshore water is shallow and this may help in the
formation of barrier beaches and other deposition
features.

A particular feature associated with the emergence of a coast is the raised beach (Fig. 10.19).
This is a beach that now stands above the present
shoreline, often appearing like an eroded rock
platform. Identification of raised beaches is best
determined from remnants of the original beach
deposits and the existence of a former cliff, especially if there are old caves in the cliff. In some
areas raised beaches are preserved virtually intact,
while elsewhere they have been largely destroyed by
later erosion. Along many coasts, several beach
levels have been recorded, varying from over 200 m
to just above present high tide. Sometimes it is
possible for levels to be correlated with identical
levels along other coasts. But it should be noted that
the same height is not always evidence of the same
age, for many areas have experienced considerable
coastal warping, which means that remnants of the
same beach may now occur at several different
heights.

### The ria coast of Guinea Bissau and Sierra Leone
Much of the West African coast between Gambia
and Sierra Leone shows typical features of drowning (World 1:250,000 sheets NC-28-2, NC-28-16).
This is largely due to eustatic change, for it has been
estimated that sea level has risen about 100 m during
the last 20,000 years. Sierra Leone has some particularly fine examples of rias, such as the Ribi (Plate
10.15), and those around the Sierra Leone River
mouth (Fig. 3.20).

Several long rias mark the coast of Guinea
Bissau, and the 60 or so islands that make up the
Bissagos Archipelago are also due to drowning

Plate 10.15 Vertical view of the drowned mouth of the river Ribi, a ria south of the Sierra Leone Peninsula

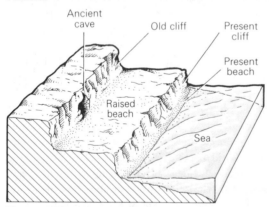

Fig. 10.19 Diagram of a raised beach

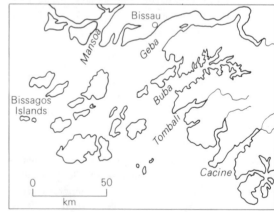

Fig. 10.20 The drowned coast of Guinea Bissau

(Fig. 10.20). It is a low region and few of the rias are deep or steep-sided, while most of the coast is bordered by an almost continuous zone of marshes and mangrove swamps.

Other examples occur on the NW Madagascar coast between Cape Ambre and Cape St Andre (the Loza River mouth is a good example of a ria); while several drowned river valleys occur on the East African coast, notably those which form the harbours at Mombasa and Dar es Salaam.

## The coastal plain of Mozambique

The flat, low-lying coast plain of Mozambique extends for more than 1500 km from near Maputo in the south to Nacala in the north. It is at its widest south of Beira, between the Save and Limpopo Rivers, where it reaches inland for 300 km. Except for occasional low hills, much of the plain is marshy and below 150 m in height. The very flat nature of the region was made apparent in January 1971, when Hurricane Felice caused the flooding of over 40,000 sq km around Quelimane, north of Beira.

The plain has resulted from the emergence of part of the continental shelf due to gradual uplift. In some areas, now inland from the coast, sand deposits show the position of former beaches and point to the stages of emergence.

Although emergence is the dominant characteristic of this coast the recent world-wide rise in sea level has drowned the lower part of the rivers, giving a submerged form to the coastline itself. Across the mouths of the rivers and embayments deposition and longshore drift have formed barrier islands, such as Bazaruto Island, and north-trending spits, such as the Inhaca Peninsula at Maputo.

## Raised beaches in Sierra Leone and Ghana

Raised beaches have been identified around the edges of the Sierra Leone peninsula at two main levels (Sierra Leone 1:50,000 sheet 61, 1:10,000 sheets 9 and 10). The Upper Beach, which is the most widespread, is at about 40 m to 50 m above sea level and slopes towards the coast at approximately 3° (Fig. 10.21). The Lower Beach occurs at about 10 m to 12 m above sea level, and is largely confined to the SE, near Hastings and Waterloo. The beaches are cut into the Bullom Series, a narrow belt of Tertiary and Quaternary sediments that surrounds the Peninsula. The beaches are due to the Quaternary uplift of the Peninsula, and their landward side is generally backed by steep slopes. In most areas they show little sign of erosion, except where streams flowing to the sea have cut narrow ravines across them. Such is the case near Hastings where the Orugu, Maroon and other rivers have dissected the Upper Beach into a number of separate zones. Many of the chief towns are sited on the Upper Beach level, for example, Freetown, Hastings and Waterloo.

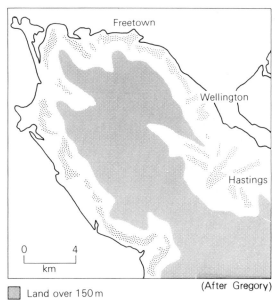

Land over 150 m

Upper raised beach

Fig. 10.21 Raised beaches on the Sierra Leone Peninsula

(After Gregory)

Several raised beaches exist along the Ghana coast, notably near Senya Beraku, 35 km west of Accra (Ghana 1:62,500 sheet 33). Much of Senya Beraku itself is sited on deposits of ancient beach pebbles at about 18 m to 20 m above sea level. Similar deposits can also be recognised at this level in a number of other places eastwards towards Fete. More recent in age are the lower beaches at 1 m and 3 m above sea level. These are particularly well seen west of Senya Beraku (Plate 10.16). The 1 m level now forms a low terrace covered by a small degraded cliff. At the base of the cliff is an old cave largely concealed by vegetation.

Other examples are: *South Africa* – A large number occur on the south coast, varying between 3 m and 30 m above sea level, such as at Hermanus, 18 m above sea level (Plate 1.6). Many beaches are backed by old cliffs with former sea caves, like Tunnel Cave at Mossel Bay. *North Africa* – Many occur along the Moroccan and Algerian coasts, for example, north of Cape Cantin on the Atlantic coast of Morocco, where five beaches are well preserved at successive levels above present high tide: 2–5 m, 15–20 m, 30–35 m, 55–60 m and 80–100 m. *West Africa* – Numerous levels have been recorded,

Plate 10.16 The 1 m high raised beach about a kilometre west of Senya Beraku, Ghana

varying from 1 m to 50 m above sea level, of which examples include 12 m at Accra, near the Anglican Cathedral, and 2 m near Nouakchott. *East Africa –* Several levels up to 90 m above sea level exist along the Kenya and Tanzania coasts, for example, in the Bagamoyo area where three beach levels can be identified at 2 m, 8 m and 12 m above sea level. Kaole, 5 km SE of Bagamoyo, is sited on the lowest of these levels.

*Plettenberg Bay, South Africa: an example of a multi-featured coastline*
Plettenberg Bay on the south coast of South Africa (Fig. 2.22) provides an example of a coast region showing a considerable variety of features. The bay extends in an arcuate form from the rocky headland of Robberg to the cliffs east of Keurboomstrand (South Africa 1:50,000 sheet 3423 AB, Fig. 10.22). The smooth curve of the bay (Plate 10.17) is mainly due to the formation of two large barrier beaches,

converging respectively from Keurboomstrand and Robberg towards the headland on which the town of Plettenberg Bay is sited. These bars were formed in front of the former coastline, which was low-lying and marshy, due to the submergence of the lower reaches of the Rivers Keurbooms, Bietou and Piesang. The northern of the two barriers, in partly closing off the mouths of the Keurbooms and Bietou Rivers, has formed a large lagoon. Along the upper part of the barriers are extensive coastal dunes, while on the modern beaches there is a well-formed system of beach cusps.

Robberg, on the south of the bay, is an unusual elongated headland rising to nearly 150 m. Off its extremity at Cape Seal lies the isolated stack of Whale Rock, while on the south of Robberg a former island has been tied to the headland by a small tombolo. At various places around the rocky sides of Robberg there are ancient sea caves, now raised about 18 m above sea level.

Plate 10.17 Plettenberg Bay, South Africa, from near Lookout Rocks towards the northern barrier beach, with the lagoon at the mouth of the Keurbooms river. In the background are the Outeniquas Mountains, part of the Cape Ranges

Plate 10.18 Cathedral Rock, a natural arch near Keurboomstrand, on the north side of Plettenberg Bay

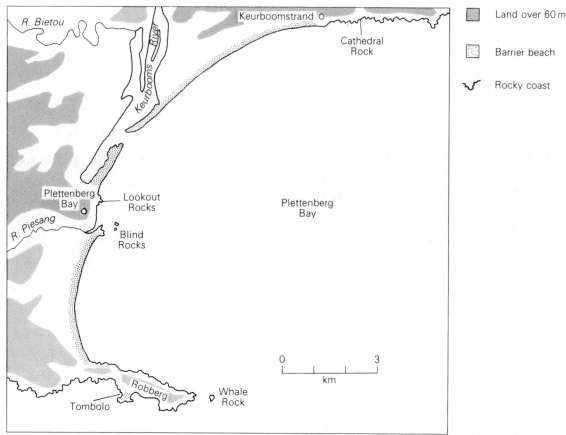

Fig. 10.22 Plettenberg Bay, South Africa

The small headland at Plettenberg Bay town is also rocky, for example Lookout Rocks and the offshore Blind Rocks, while along the cliffs east of Keurboomstrand, Cathedral Rock is a good example of a natural arch (Plate 10.18).

## Coral coasts

Coral coasts differ from most other coasts since they are largely organic in origin. They form from small sea organisms called coral polyps, which tend to live in colonies. When polyps die their skeletons, which are made of calcium carbonate, accumulate with other organisms to form coral limestone. In this way large banks of rock are gradually built up called coral reefs.

Corals thrive best in warm clear, salt water, with a temperature of 20°C to 30°C, and a salinity of 27 to 40 parts per thousand. Both light and constantly moving water are essential, hence they are mostly found between low tide level and a depth of 30 m. For these reasons reefs are mainly concentrated on the west side of oceans, away from cool currents, between 30° north and south of the equator. A further limiting factor, and one causing the virtual absence of coral along the west coast of Africa, is extensive longshore movement of sand. Not only does sand smother young coral, but also it fails to provide a firm base for reef growth.

Many reefs are extremely old, dating back to the Tertiary era, and some of these are dead and no longer contain living coral. This is often because they are now permanently above low tide level, or have been submerged to great depths, both due to relative changes in sea level (page 216).

**Types of reef:**

1  *Fringing* – a coral platform, up to 1 km wide, joined to the coast, or separated from it by a shallow lagoon.

2  *Barrier* – a coral platform separated from the coast by a wide, deep lagoon. It may form off a mainland coast and also as a ring round an island.

3  *Atoll* – a ring of coral surrounding a fairly deep lagoon, but generally broken in places by narrow channels.

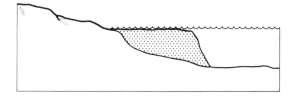

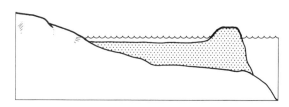

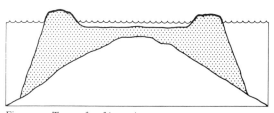

Fig. 10.23  Types of reef (1, 2, 3)

Fringing reefs and some barrier reefs form by gradually building up from the floor of the shallow offshore zone till they reach sea level. But this does not explain the formation of atolls and many barrier reefs that originate from depths far greater than coral can survive, often in excess of 1000 m. Recent seismic exploration and drilling, however, has given weight to the theory first expounded by Charles Darwin in 1842, that atolls and island barrier reefs began life on the edges of volcanic islands.

When a volcanic island first erupts its weight will destroy the isostatic equilibrium of the crust in that area (page 8). In time, due to isostatic adjustment, the volcano will begin to sink under its own weight, and, according to the size of the volcano, the crust may subside several hundred metres or more. If coral had begun to grow as a fringing reef at sea level on the flanks of the volcano (Fig. 10.24(A)), the first effect of isostatic sinking would cause the coral to transform to a barrier reef, providing the coral growth could keep pace with the sinking (Fig. 10.24(B)). Later, if the volcanic island became completely submerged, the barrier reef would become an atoll (Fig. 10.24(C)). Coral reefs grow very slowly, and atoll formation depends on the coral being able to grow at a faster rate than the volcano is sinking. Those that cannot keep pace disappear under water to become seamounts. However, it is probable that other factors, such as wave movement in the open ocean and sea level changes, have also played a part in reef formation.

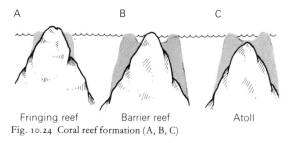

Fig. 10.24  Coral reef formation (A, B, C)

*The fringing coral reef of the East African coast*
A fringing reef lies off much of the Kenya and Tanzania coast. It continues north to Somali and south to Mozambique (Kenya 1:50,000 sheets 198/2, 201/1). In places, such as south of Kilifi, the reef is close to the shore, but mostly it is about 500 to 2000 metres from the shore. The distance would seem to be related to the growth of certain plants and deposition of sediments, both of which tend to impede coral formation. The low salinity of the water opposite river mouths also affects coral, and there is a marked gap in the reef at the mouths of

Plate 10.19  View at low tide of the undercut raised coral platform on the north side of Oyster Bay, Dar es Salaam

muddy rivers and creeks, for example at Mombasa, Kilifi and Takaungu. The reef's seaward edge is steep due to the more active development of the coral here, where the constantly moving water of the breaking waves brings in a regular supply of oxygen and plankton for coral growth.

Many parts of the mainland coast are themselves built of ancient raised coral reefs. These have been eroded into low cliffs, fronted by a wide platform on which are numerous flat-bottomed pools. The cliff has an undercut notch, which in places is enlarged into caves, caused by the extensive dissolution of the limestone (Plate 10.19). The overhang above the notch, the visor, is a zone of sharp, jagged ridges and pinnacles, known as coastal lapiés. These lapiés result from the action of sea spray and rain solution. All these features can be seen at Oyster Bay, north of Dar es Salaam. Other examples are: the Red Sea coast; the west coast of Madagascar and the Mauritius coast.

*The barrier reef of Mayotte, Comoro Islands*
Mayotte is the southern-most of the four Comoro Islands, at the north end of the Mozambique Channel (Fig. 2.4). It is volcanic in origin and rises from the ocean floor (Mayotte 1:50,000 IGN). Mayotte has been severely dissected by erosion and its irregular coastline with numerous bays and inlets is strong evidence of drowning. The coast is bordered by a narrow fringing reef, but more significant is the extensive barrier reef that almost completely surrounds Mayotte (Fig. 10.25). The lagoon between the island and the barrier reef is deep and up to 10 km wide in places. In the south, off Kani-Kélé there is even a double barrier reef.

A barrier reef has also formed off parts of the west

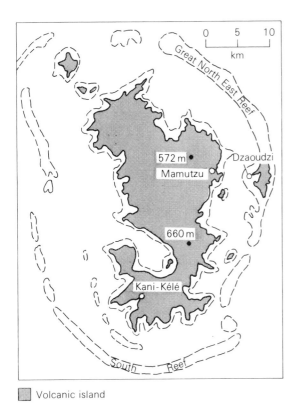

Volcanic island

~- Coral reef

Fig. 10.25 Barrier reef of Mayotte

coast of Madagascar between Morombe and Tulear, notably the Grand Reef off Tulear.

*Aldabra Atoll*

Aldabra is a large atoll that lies some 700 km off the coast of East Africa. It is one of more than 100 islands that extend north-east towards the Seychelles (Aldabra 1:25,000 DOS). Aldabra is about 34 km long and up to 14 km wide. It is composed of three main islands separated by narrow channels leading to the central lagoon (Fig. 10.26). It is a raised atoll, much of it now being several metres above high water level, with a sharp-edged, pitted and jagged rock surface known as 'champignon'. The seaward edge of the reef is very steep, but on the other side there is a gentle slope to the shallow lagoon. The north coast has been undercut and often forms low cliffs, but much of the south coast has a beach with coastal dunes in places. The greater part of the atoll surface is covered by thick grasses and thorn bushes, with mangrove and other swamps around the edge of the lagoon.

Not all coral islands are true atolls. Europa Isle, for example, 350 km NE of Tulear, Madagascar, is an ancient atoll on which most of the coral is now dead and the lagoon has been silted in to form a central plain (Plate 10.20).

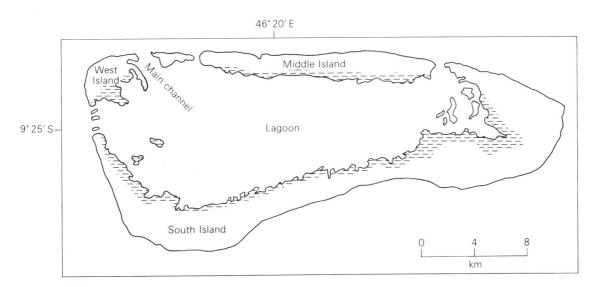

--- Mangrove

Fig. 10.26 Aldabra Atoll

Many of the islands north-east of Aldabra, such as Desroches (53°40'E, 5°42'S) in the Amirante Islands, are sand cays. A sand cay is a sand bank that forms an island on top of a coral reef.

Other examples are: Astove (10°7'S, 47°46'E) and Cosmoledo (9°42'S, 47°35'E) both coral atolls near Aldabra (Fig. 2.4). Bird Island (3°55'S, 55°12'E) in the Seychelles is the only one of these islands that is an atoll, the others being composed of granite rock (see page 10).

## Classification of coasts and islands

The great variety in the nature of coasts makes classification difficult. The problem has been complicated by the large number of sea level changes that occurred during the last two million years. The simplified classification given here is based on that by the German author H. Valentin (summarised in Cotton C. A., 1954, *Geographical Journal*, Vol. 120, pages 353–361).

The classification may be justified in that it attempts to include the majority of coastal forms, while at the same time remaining simple. However, this overlooks the fact that few coasts have a simple nature. Also, since world sea level has risen about 100 m during the last 20,000 years, all coasts could arguably be classified as drowned. Many parts of the African coast provide examples of several of the groupings in the classification, as was illustrated on pages 220–222.

### Classification of islands

Islands vary in size from huge features like Madagascar, fourth largest island in the world, to the miniature Annobon off the Gabon coast. Some occur individually, while others are found in a group known as an archipelago, for example the Canary Islands. The following classification is based on origin.

### Classification of Coasts

| | EXAMPLES |
|---|---|
| 1 *Coasts of Retreat and Submergence* | |
| (a) Due to erosion: | |
| Cliffed coasts | Algeria (page 204) |
| (b) Due to submergence: | |
| Non-glacial, rias | Guinea Bissau (page 218) |
| Glacial, fiords | Chile, S. America (page 179) |
| 2 *Coasts of Advance and Emergence* | |
| (a) Due to inorganic deposition: | |
| (i) Marine – spits and barrier beaches | W. Africa (pages 208, 212) |
| (ii) Fluvial – delta coasts | Medjerda, Tunisia (page 213) |
| (iii) Volcanic – eruption of lavas | Cabo de San Juan, Canaries (page 38) |
| (b) Due to organic deposition: | |
| (i) Coral coasts | Kenya (page 223) |
| (ii) Mangrove coasts | Nigeria (page 216) |
| (c) Due to emergence | |
| (i) Coastal Plains | Mozambique (page 217) |
| (ii) Raised Beaches | Sierra Leone (page 219) |

### Islands

| Origin | Example |
|---|---|
| Marine Erosion | Goree and Madeleines (page 202) |
| Marine Deposition | Bazaruto (page 219) |
| Volcanic Eruption | Mayotte (page 225) and Reunion |
| Coral | Aldabra (page 225) |
| Coastal Submergence | Bissagos Archipelago (page 218) |
| Continental Drift | Seychelles (page 10) |

Plate 10.20 Part of the coral island of Europa, at the south end of the Mozambique Channel

## References

BATTISTINI R., 'Les Caractères Morphologiques du Secteur Littoral compris entre Foulpointe et Maroantsetra', *Madagascar Revue de Géographie*, No. 4, pp. 5–36, 1964.

GALLIERS J. A., 'Barrier Beaches and Lagoons of the Ghana Coast', in British Geomorphological Research Group Occasional Paper No. 5 'Geomorphology in a Tropical Environment', 1968.

GREGORY S., 'The Raised Beaches of the Peninsula Area of Sierra Leone', *Transactions of the Institute of British Geographers*, No. 31, pp. 15–22, 1962.

GUILCHER A. and NICOLAS J. P., 'Observations sur la Langue de Barbarie et les bras du Sénégal aux environs de Saint Louis', *Bulletin d'Information du Comité Central d'Océanographie et d'Etude des Côtes*, Vol. 6, pp. 227–242, 1954.

MCINTIRE W. G. and WALKER H. J., 'Tropical Cyclones and Coastal Morphology in Mauritius', *Annals of the Association of American Geographers*, Vol. 54, pp. 582–596, 1964.

PUGH J. C., 'A Classification of the Nigerian Coastline', *Journal of the West African Science Association*, Vol. 1, No. 1, pp. 1–12, 1954.

STODDART D. R., 'The Conservation of Aldabra', *Geographical Journal*, Vol. 134, pp. 471–486, 1968.

TEMPLE P. H., 'Aspects of the Geomorphology of the Dar es Salaam Area', *Tanzania Notes and Records*, No. 71, pp. 21–54, 1970.

TRICART J., 'Aspects et Problèmes Géomorphologiques du Littoral Occidental de la Côte d'Ivoire', *Bulletin de l'Institut Français d'Afrique Noire*, Vol. 19, Série A, pp. 1–20, 1957.

USOROH E. J., 'Recent Rates of Shoreline Retreat at Victoria Beach, Lagos', *Nigerian Geographical Journal*, Vol. 14, No. 1, pp. 49–58, 1971.

# CHAPTER ELEVEN

# Lake basins

A lake is a body of water contained within a basin or hollow. The size, depth, and permanence of a lake largely depends on the nature of its basin. For this reason, a classification of lakes is most suitably based on the origin of lake basins. Broadly speaking, there are 2 main groups of lake basin. First, those that result from a hollow in the earths surface formed by erosion, vulcanicity or earth movements. Second, those formed by a barrier of material deposited, for example, by ice, water or volcanic activity.

By far the most common cause of lake formation is glaciation. But not all lake basins are the result of a single cause, and two factors at least are often involved. In glacial areas, the depth of many lakes is due partly to the erosion of glacial troughs and partly to a moraine ridge barrier. The effect of the barrier is to raise the lake level and sometimes increase the lake area.

The size and permanence of a lake also depends on the inflow of water from rainfall and rivers, relative to the outflow by evaporation and rivers.

Confusion can arise over the definition of a lake, since some very large lakes are called seas. The Caspian Sea is the world's largest lake, with Lake Superior second and Lake Victoria third.

Plate 11.1 Rock basin lakes in the upper Hobley Valley, Mt Kenya

Several African lakes show great seasonal fluctuations in size, while the flood plain and delta lakes of river valleys are often very impermanent. There are extreme seasonal variations in the area of Lake Chad, 12,000 to 22,000 sq km, which is the shallowest of Africa's major lakes and yet, at its maximum, one of the largest (page 241). Lake Chad (Plate 11.5) only 7 m deep, is in sharp contrast to Lake Tanganyika which at 1425 m deep is among the world's deepest and half of it is below sea level.

## Origins of major types of lake basin

### 1. Glaciation

*Rock basin lakes*
These lakes lie in basins either formed or deepened by glacial erosion, and include cirque lakes (page 177), glacial trough lakes (page 179), lakes in over-deepened passes, and rock basin lakes on ice-eroded plains (page 175).

Several cirque lakes occur on Mount Kenya (Kenya 1:25,000 Mt Kenya Special Sheet), for example Teleki Tarn (page 183), Hanging Tarn (Plate 9.3), Gallery Tarn and Hidden Tarn.

On Ruwenzori (Uganda 1:25,000 Ruwenzori Special Sheet), Lac du Speke and Lac Catherine are cirque lakes. Lac du Speke (Fig. 11.1) lies in a small steep-sided cirque at about 4200 m on the west slope of Mount Speke. Two islands rise above the lake surface, and the rock walls around the lake are vertical for the first 20 m.

The Carr Lakes and the Enchanted Lakes on Mount Kenya all lie in valley rock basins (Plate 11.1), and so does Lake Michaelson (Plate 9.3). Similar rock basin lakes on Ruwenzori are Lac Vert (Plate 9.1) and Lac Noir (Plate 11.2), both in the steep upper valley of the Kamusoso (Fig. 9.16). The outflow from the narrow, elongated Lac Noir is by underground seepage. Another rock basin lake on Ruwenzori is Lac de la Lune on the north side of the Roccati Pass (Fig. 9.14).

*Moraine ridge barrier lakes*
Many glacial lakes occupy basins formed by the damming of a valley by a terminal moraine ridge (page 176). On Mount Kenya, Tyndall Tarn at the

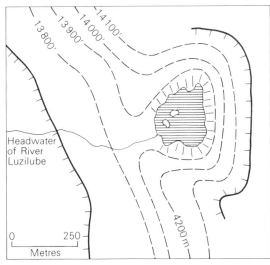

Fig. 11.1 Lac du Speke, Ruwenzori

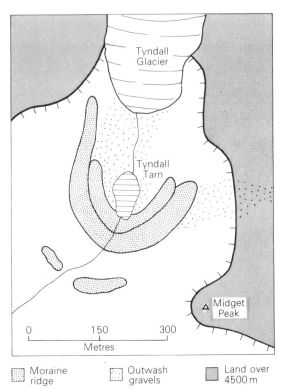

Fig. 11.2 Tyndall Tarn, Mt. Kenya

Plate 11.2  Lac Noir, Ruwenzori, viewed from the NW

Plate 11.3 Hut Tarn, sited on a col above the upper Teleki valley, Mt Kenya. Beyond is the rock pinnacle of Point John, with the Darwin Glacier to the left. Part of the lateral moraine damming the lake is visible to the right

head of the Teleki Valley is a moraine-dammed lake (Figs. 11.2 and 9.15). The lake is only about 100 m long, and lies between the moraine barrier and the snout of the Tyndall Glacier, from which it is fed by a small melt-water stream. It is one of the most recent Mount Kenya lakes and was not in existence when the Tyndall Glacier was sketched by J. W. Gregory in 1893. The moraine barrier is in the form of two arcuate ridges, both about 30 m wide.

Other moraine-dammed lakes include Lac Gris on Ruwenzori (Fig. 9.16), Lake Hohnel and Hut Tarn (Plate 11.3) on Mount Kenya. The last two are initially rock basin lakes, Lake Hohnel occupying a cirque, but their size has been considerably increased by the moraine barriers (page 184).

*Lakes occupying depressions in glacial drift*
Kettle lakes and lakes in depressions caused by the irregular deposition of glacial shift are most common in areas of extensive deposition by ice sheets, (page 176). For this reason such lakes are not to be expected in Africa. However, they may sometimes

Plate 11.4 The shore of Lake Malawi, near Salima. This is the second largest of the rift lakes, covering about 30,000 sq km. It has a maximum depth of 706 m

develop in drift deposits on the floor of glacial troughs. Lake Mahoma in the lower Mubuku Valley of Ruwenzori is a kettle lake in the lateral moraine ridge at the edge of the glacial trough. It is about 300 m in diameter and less than 10 m deep, lying about 1¼ km SW of Nyabitaba (Nyinabita) Hut (page 187).

## 2. Earth movements

### Rift valley lakes

An elongated outline, steep sides and a great depth are the significant characteristics of these lakes. The main examples in the world are to be found in the rifts of East Africa, notably Lakes Tanganyika, Malaŵi (Plate 11.4) and Turkana (Rudolf). Smaller ones include Lakes Rukwa and Eyasi (Fig. 2.4). The great length of Lake Tanganyika (680 km) and its narrow width (50 km) are a close reflection of the shape of the rift trough. The lakes lie on the floor of the rifts in basins which are largely the result of differential movements associated with faulting across the valleys, causing some parts of the floor to be downfaulted lower than other parts. In some places the long sides of the lakes are actually bordered by fault scarps (pages 16–17), for example the east side of Lake Mobutu north of Tonya and the west side of Lake Malawi between Florence Bay and Nkhata Bay. But more often there are varying widths of sedimentary deposits between the fault boundaries and the lake shores (Plate 2.3). The inflow and outflow of rift lakes tends to be at their narrow extremities, but this is not always the case and Lake Tanganyika drains west via the Lukuga to the Congo system.

As well as faulting, infilling of sediments and extensive lava flows (page 17) have also influenced the shape of several of the present basins. In fact, many of the smaller lakes in East Africa (Fig. 2.5) although lying in the rifts, occupy basins that result more from barrier formation than faulting, and for this reason are often shallow.

### Lakes due to crustal warping

These lakes are generally large in area and very shallow. Examples include Lake Victoria (Fig. 2.11), Lake Chad (Plate 11.5), and Lake Bangweulu,

in Zambia. Lake Victoria, Africa's biggest lake, covers almost 69,000 sq km and lies in a broad, shallow, downwarped basin between the East and West Rifts (Fig. 2.7). The downwarping was followed by upwarping to the west which caused rivers like the Kagera to reverse their direction and run backwards into the crustal depression and form Lake Victoria. The maximum depth of the lake is only 82 m, and the shoreline shows the influence of drowning in the form of numerous inlets and islands.

In North Africa, many of the Algerian and Tunisian chotts (page 29) lie in shallow basins caused by local downwarping within the Atlas Mountains, for example Chott ech Chergui. These chotts form a chain of muddy saline depressions that are generally occupied by water only during winter.

## 3. Vulcanicity

### Lakes occupying volcanic calderas and explosion craters

Calderas and explosion craters (pages 37–38) generally form roughly circular, steep-sided lake basins. A typical caldera lake is Ngozi (Fig. 11.3), 17 km SE

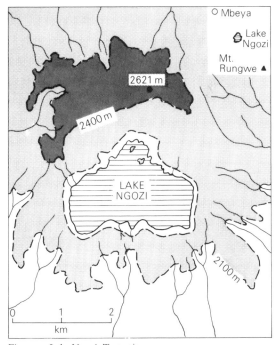

Fig. 11.3 Lake Ngozi, Tanzania

233

Plate 11.5  Part of the vast expanse of Lake Chad with its numerous reed-covered islands

of Mbeya, Tanzania (Tanzania 1:50,000 sheets 245/3 & 259/1). The original volcano was built of trachyte and layers of this lava are visible in the steep caldera walls around the lake. Ngozi is more than 75 m deep and the two small islands near the north shore may be the peaks of extinct secondary cones. By far the largest caldera lake in Africa, however, is Lake Shala, in the Ethiopian rift. It has an elliptical shape with a width of over 25 km. Surrounding the deep waters (250 m) are vertical walls rising above 150 m (Fig. 2.4).

Other examples of caldera lakes include Lakes Wum, Nyos and Oku, north of Bamenda in Cameroon (Fig. 5.36); Lake Tritriva near Antsirabe in Madagascar (Plate 11.6); and Grand Bassin, 11 km south of Curepipe, Mauritius.

Lakes also form in the craters of extinct or dormant volcanoes, such as Muhavura (page 44). They are not common in ash and cinder cones, since the pyroclasts are generally too permeable to hold water. The walls of explosion craters, however, are partly made of local rock, hence many of these craters hold lakes. Numerous examples occur in SW Uganda (Fig. 3.6), notably Lake Nyungu (Plate 3.3).

### Lava-dammed lakes

Multiple lava flows often play a major role in drainage diversion. The eruption of the Virunga

Plate 11.6  Lake Tritriva, 12 km SW of Antsirabe, Madagascar, is a perfect example of a lake lying in the bottom of a steep-sided caldera. Similar features nearby are Lakes Tritrivakely and Andraikiba

volcanoes (page 41) blocked the West Rift and caused the formation of Lake Kivu. Individual lava flows may dam a river valley and pond back the waters to form a lake the same shape as the valley, for example Lakes Mutanda (Plate 11.7), and Kayumba in Uganda; Lakes Bulera and Ruhondo in Rwanda; and the Mokoto Lakes in Zaire (Fig. 3.9).

Lake Mutanda (Fig. 11.4) lies amongst the Busanza Hills of SW Uganda. Its outlet is the Kaku River, and so into the Rutshuru. The lake has a maximum depth of 50 m, and was dammed by a series of lava flows from the small cones south of Kisoro, in particular Sagitwe (Uganda 1:50,000 sheets 93/1 & 93/3).

In central Madagascar (Fig. 3.10) Lake Itasy was formed by lava blocking the Matiandrano River and its tributaries (Madagascar 1:100,000 sheet M.47). Originally the lake (Plate 3.6) was 25 m higher than at present, but it is being steadily filled in by alluvium, while its outlet, the River Lily, has cut a narrow gorge 20 m deep and drops over a small waterfall (Plate 5.19). Several smaller barrier lakes lie west of Lake Itasy.

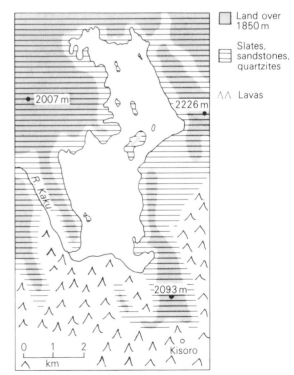

Fig. 11.4  Lake Mutanda, Uganda

Plate 11.7 Lake Mutanda, a lava-dammed lake on the SW Uganda border. In the background from left to right are the volcanoes of Muhavura, Mgahinga and Sabinio (Fig. 3.9)

## 4. Other processes

*Lakes dammed by landslides and other waste debris*
Many small lakes form from the mass movement of waste debris across river valleys. Such lakes, however, are often short-lived.

A temporary lake formed in the Mbaka Valley, Tanzania in 1955 as a result of a landslide (page 72), but it broke through the barrier and drained away after eight hours. It was a much larger landslide that blocked the course of the Mutale River in the Soutpansberg, South Africa, and formed Lake Funduzi. This lake is about $1\frac{1}{4}$ km long and is bordered by steep scarps to the north and south (Fig. 11.5 and South Africa 1:50,000 sheet 2230 CD). Water from the lake now drains into the Mutale by seepage through the sandstone rubble of

the landslide. Lake Bujuku, Ruwenzori (Fig. 9.14), may also be due to an old landslide.

Several lake basins have been formed by material washed down from the gullies of the Awku Uplands, SE Nigeria, especially Lakes Agulu and Iyiocha (Fig. 4.6 and Nigeria 1:50,000 sheet 301 SW).

In SW Uganda, Lake Nyabihoko (Karengye) was formed by the deposition of alluvial fans at either end. The lake is situated on the swampy watershed between drainage flowing to Lake Amin and that flowing to Lake Victoria. The water from Lake Nyabihoko drains via the Rufua River through swamps to the Kagera (Fig. 2.11 and Uganda 1:50,000 sheet 85/3).

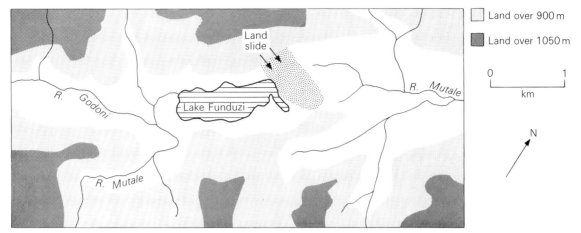

Fig. 11.5  Lake Funduzi, Transvaal, South Africa

## Flood plain and delta lakes

This group includes all those that result from deposition along the lower course of river valleys. They are very shallow and vary greatly in size, some being large while others are only a few metres long. They are extremely numerous, and, since most are temporary, they rarely have individual names. Some form in deltas and flood plains due to levees which prevent flood water returning to the river channel (page 87). An example is Lake Matshi in Zaire, one of a number that have formed behind levees and are now cut off from the Loange River, a tributary of the Congo. Lake Matshi is about $3\frac{1}{2}$ km long and lies 150 km east of Kitwit. During floods it is replenished with water, but in the dry season it is less than 1 m deep. Several lakes have formed in the inner part of the Ogooué delta in Gabon (Fig. 3.24) for example Lakes Avanga and Ezanga.

West of Dar es Salaam (Tanzania 1:50,000 sheet 186/3) a group of small lakes have developed at the lower end of valleys tributary to the Sinza and Luhanga Rivers, that flow into Msimbazi Creek. The lakes, notably Magomeni and Mwananyamara, have formed due to extensive alluvial barriers deposited by the main streams across the mouths of the tributaries.

A common type of flood plain lake is the oxbow (Fig. 11.6 and page 87). They are very numerous in many valleys, for example the Tana in Kenya, near Garissa (Plate 11.8, Fig. 5.11, and Kenya 1:100,000 sheet 140); the Galma in Nigeria near Zaria (Nigeria 1:50,000 sheet 102 SE); and the Wilge in South

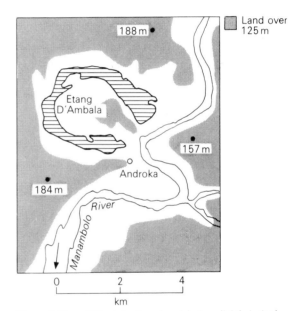

Fig. 11.6  Etang D'Ambala. An oxbow lake in a slightly incised abandoned meander of the Manambolo River, about 70 km SE of Antsalova, Madagascar

Africa at Harrismith (South Africa 1:50,000 sheet 2829 AC).

## Coastal lagoons

Lagoons are often described as lakes. They tend to be shallow, and originate from the growth of a barrier beach across a river mouth, so forming a barrier to the outflow of water (page 202). The West African coast east of Cape Palmas has a large

237

Plate 11.8 Oxbow lakes in the Tana river valley, SW of Garissa, Kenya

number of lagoons, for example Lake Nokoué in Benin (Fig. 10.10), Korle Lagoon at Accra, and Ebrié Lagoon in Ivory Coast.

The lagoons are the result of longshore drift building sandbars along a submerged coast and enclosing the drowned parts of river valleys. Similar lagoons to the West African ones occur on the Gabon coast, for example Lake Igéla, and the east Madagascar coast.

### Lake basins due to wind deflation

Basins or blowouts excavated by wind deflation to the depth of the water table (page 158–9), may hold permanent lakes and give rise to oases in desert areas. Examples include Birket Siwa and Birket Zeitun in the Siwa depression, west of Qattara (Fig. 8.4) in NW Egypt (birket = lake). The small Maghra Lake (Fig. 8.14) on the floor of Quattara is a similar feature.

### Lake basins due to solution

In karst regions, lake basins form from the effects of solution and collapse. The depth of such lakes is determined by the level of the water table (page 126). Collapse dolines may hold lakes where collapse is into water-filled caves, for example Lake Otjikoto 20 km NW of Tsumeb, in the Otavi dolomite region of Namibia (Fig. 8.2 and Plate

Plate 11.9 Lake Ojikoto, a circular lake about 180 m diameter, lies in the cauldron-like depression of a collapse doline near Tsumeb, Namibia

11.9). It is about 180 m deep, but this varies with seasonal changes in the water table. Nearby is a smaller but similar depression holding Lake Guinas. Solution lakes also occur in poljes, where the basin is partly caused by solution and partly by tectonic activity. Many of the seasonal and permanent lakes of the Moroccan Middle Atlas lie in poljes and large dolines. They are known by the Berber name of aguelman, for example, Azigza Aguelman and Afriroua Aguelman (page 128).

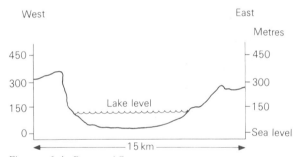

Fig. 11.7 Lake Bosumtwi Crater

### Meteorite crater lakes

This type of lake basin is relatively rare compared to volcanic craters. The crater results from the impact of a meteorite from space. One of the largest meteorite crater lakes in the world is Bosumtwi, SE of Kumasi in Ghana, that lies in a well-preserved crater of over 10 km diameter (Fig. 11.7). The evidence for meteorite origin at Bosumtwi includes plucking marks on the crater walls and the occurrence of coesite, a mineral known to form only at very high pressure and generally associated with meteorites (Ghana 1:62,500 sheets 91 & 129).

### Man-made lake basins

These lakes are formed by the construction of a dam across a narrow, steep-sided section of a river valley, often a gorge, such as at Akosombo on the Volta (Plate 11.10) and at Kariba on the Zambesi. The river then ponds back flooding low-lying areas upstream from the dam. Other man-made African lakes include Kainji and Nasser (Fig. 8.4).

This category also includes the numerous small reservoirs constructed by farmers in parts of the savannas, together with those built in mining

Plate 11.10 The 113 m high rock-fill dam at Akosombo in the Volta gorge of southern Ghana, which holds back the largest artificial lake in the world, 8,482 sq km

districts like the Jos Plateau, especially east of Bukuru (Nigeria 1:50,000 sheet 168 NE).

Many lakes, especially the large ones, have examples of shoreline features similar to those that develop along sea coasts, for example, raised beaches and sand spits. On the south shore of Lake Malaŵi, north of Mangochi (Fort Johnston), there are several raised beach levels, some of which are about 45 m above the present lake. Similar features can be identified along the north shore of Lake Victoria.

The best examples of spits are on the Uganda side of Lake Mobutu, where there are 9 main ones, especially those at Tonya and Kabira Point, approximately 25–30 km north of Butiaba (Fig. 2.11).

In the context of geological time, even the largest lakes are only temporary features. If the water balance is not maintained the lake will drain away or evaporate. Alternatively, the lake may become filled with sediments, or the outlet may be lowered by progressive erosion. All over Africa there is evidence of former lakes, such as strandlines, for example along the edge of the Makarikari Depression, Botswana (pages 168–69); marshy depressions, for example Ifanja Marsh, 15 km north of Lake Itasy, Madagascar (Fig. 3.10); and recent lake sediments, for example the Suguta Valley, south of Lake Turkana (Fig. 3.7).

Lakes Naivasha, Elementeita, and Nakuru in Kenya (Fig. 2.5) were once part of a much larger lake during the Pleistocene. At the south end of Lake Naivasha lake waters cut the deep Njorowa Gorge as an overflow channel to the Suswa Plains. Former lakes also once occupied the Congo Basin and the area of the Niger Inland Delta (page 94). The Congo gorge below Kinshasa is the overflow channel that drained the former lake (page 113).

Lake Chad (Plate 11.5) still remains one of Africa's largest lakes, but in Pleistocene times it was over 300,000 sq km, about 15 times its present size, while its depth must have been at least 150 m. Evidence of this is seen in shore lines many metres higher than the present lake level, notably the Bama sand ridge near Maiduguri in northern Nigeria.

## References

DUCKWORTH E. H., 'Crater Lakes in Bamenda Province', *Nigeria Magazine*, No. 37, pp. 65–79, 1951.

GROVE A. T., 'Rise and Fall of Lake Chad', *Geographical Magazine*, Vol. 42, No. 6, pp. 432–439, 1970.

JANISCH E. P., 'Notes on the central part of the Zoutpansberge Range and on the origin of Lake Funduzi', *Transactions of the Geological Society of South Africa*, Vol. 34, pp. 151–162, 1931.

MACLAREN M., 'Lake Bosumtwi, Ashanti', *Geographical Journal*, Vol. 78, pp. 270–276, 1931.

MOTTRAM B. H., 'The Origin of Lake Nyabihoko, Ankole', *Uganda Journal*, Vol. 32, pp. 183–188, 1968.

SHEPHERD A., 'The Sand Spits of Lake Albert', *Zeitschrift für Geomorphologie*, Vol. 5, pp. 53–72, 1961.

TEMPLE, P. H., 'The Lakes of Uganda' in *Studies in East African Geography and Development* edited by S. H. Ominde, Heinemann, 1971.

# Index

*Figures in bold denote a plate on that page*

Aberdare Range 17, 19, 52, 172
ablation 172
abrasion 80, 81, 156, 164, 174, 192, 195–6, 198–9
accordant drainage 105–6
accordant summits 139
acid lava 34, 37
adjustment of drainage 104–5, 108, 112, 121
Ado Rock, Nigeria 66, 148, 149, 150
Adrar Massif, Mauritania 120, 153, 165, 166, 168
adventive cone, *see* parasitic cone
Afar Triangle, Ethiopia 16–17, 21, 22, 32, 36, 39, 41, 52, 74
Afrera L., Ethiopia 16, 21, 41, 52
African Surface 138, 144–6, 165
Afriroua Polje, Morocco 128, 129, 239
Agadir earthquake 33
aggradation 1
aguelman 128, 239
Agulu L., Nigeria 75, 236
Ahémé Lagoon, Benin 207–8
Air Mts. 39, 59, 152–3, 163
aklé 160
Akwapim Hills, Ghana 29, 103
Albert L., *see* Mobutu L.
Aldabra Is. 16, 225–6
alluvial cone 48, 96, 155, 166
alluvial fan 95–6, 164, 166, 178, 236
alluvial plain, *see* flood plain
alluvium 81, 85–7, 102, 108, 165, 235, 237
Alpine Mountain Building 4, 28
Ambereny Massif, Madagascar 56
Amin L. 19–21, 39, 41, 236
ancient lake basins 94, 103, 112–13, 168–9, 180–1, 240–1
ancient sand dunes 169
ancient watercourses 158, 168
Andrafiabe Cave, Madagascar 131, 132, 133
Andringitra Massif, Madagascar 120–1
Ankarana Plateau, Madagascar 131, 132–3
annular drainage 105
antecedent drainage 28, 108–9, 113, 114
anticlinal valley 29, 31, 105–6
anticline 27, 29, 31, 105, 109, 122, 202
Antongil Bay, Madagascar 16, 27
aquifer 8
arch 199–200, 204, 205, 221, 222
Archean rocks 6
arcuate delta 92–3
arête 178, 181, 183, 185, 187, 188
ash cone 37, 39–40, 234
Aswa R. 24, 27
Atakor, Hoggar Mts. 43–4, 153, 168
Atlas Mts. 4, 9, 11, 28–30, 67, 75, 109, 128–9, 150, 172, 192, 196, 233
atoll 223, 225–6
attrition 80

Augrabies Falls, S. Africa 90, 111, 116, 140
aven 129–130
Awka Uplands, Nigeria 75, 122–4, 236
Azigza aguelman, Morocco 128, 239

backslope 101, 117–18
bajada 166
Baker Mt., Ruwenzori 21, 67, 174, 184–5, 186, 188
Bamenda Scarp, Cameroon 23
Bandiagara Plateau, Mali 118, 119–120
bankfull 85
barchan 159–160, 162, 163, 166, 215–16
Baringo L., Kenya 17–18
Barotse Plain, Zambia 86, 91, 116
barrier beach 201–2, 207, 208, 209, 214–17, 220, 222, 226, 237
barrier reef 223–5
basal sapping 177
basal surface of weathering, *see* weathering front
basalt dome 37, 41, 42
basalt rocks & scenery 2, 6, 7, 17, 37, 41–2, 50–2, 64, 72, 112, 128, 190, 202, 207
base level of denudation 11–12, 103, 143, 168
base level movements 12, 81, 96–8, 100–1, 114–15, 138, 216–17
Basement rocks or complex 6, 19–20, 27, 29, 39, 47, 54, 58–9, 109, 112, 145, 146
Bashee R., S. Africa 99
basic lava 34, 37, 40, 49
basin, structural, *see* structural basin
basin and range country 16
batholith 52, 54, 59–60
Batian, Mt. Kenya 48, 173, 183, 184–5
Batoka Gorge 27, 28, 111–13
bay 195–7, 200, 204, 224
Bazaruto Is., Mozambique 219, 226
beach 92, 197–8, 200–4, 207–8, 209, 210–11, 214–16, 219, 225
beach rock 201
bedding plane 2, 3, 7, 65, 69–71, 73, 81, 125, 127, 170, 199
beheaded stream 101
Benue Rift, Nigeria 4, 20, 29, 49, 89, 106
Benue river & valley, Nigeria 58, 89, 90, 106, 138, 146
berm 201, 210
Betsiboka R., Madagascar 92–3
birdfoot delta 92, 94
Birim R., Ghana 97–8, 101
Bismark Rock, Tanzania 148
Bissagos Is., Guinea Bissau 217–18, 226
Biu Plateau, Nigeria 52
blind valley 126–7, 129
block disintegration 62, 66
block mountain 15, 17, 20–3, 135, 150
blowhole 199–200, 207
blowout 156, 158–9, 166, 238
Bogenfels, Namibia 153, 204, 205
Bolleri R., Nigeria 106

Bongo Rocks, Ghana **144**
bornhardt 140–3, 145–150, **163**, 164
Bosumtwi L., Ghana 105, 239
Boti Falls, Ghana **3**, 112
boulder clay, *see* till
bowal 68
braided river channel 83, 85, **86**, 90, 95–6, 115, 117
Brandberg, Namibia 140
breached anticline 106
breached watershed 175
Bujuku L., Ruwenzori 184, **186**, 236
Bujuku Valley, Ruwenzori 80, 184–5, **186**, 187
Bushveld Basin, S.Africa 57
butte **55**, 68, 118–120, 140, 164–5
Butiaba Scarp, Uganda 21

Calansho Serir, Libya 152, 154
caldera 38, 43, 46–7, **48**, **49**, 53, 233–5
calcite formations 128, 132, **134**, 135
calcrete 68–9, 158
Cameroon, Mt. 36, 38–9, 46, 150
Canary Is. 36, 38, 40, 46, 226
Cango Caves, S.Africa 135
canyon *see also* gorge; 113–14, 153, 161–2
Cape Manuel, Senegal **7**, 202
Cape Point, S.Africa **197**
Cape Ranges, S.Africa 4, 11, 28–30, **31**, 67, 104, 150, 172, 196, 221
Cape Verde Is. 36, 39
Cape Verde Penin., Senegal 202–3, 215
carbonation 62, 64, 67, 124–5
carbonatite magma 19–20
Carboniferous Ice Age 171–2, 189
Carr Lakes, Mt. Kenya 181, 188, **228**, 229
castle kopje 141, 143, 145–6, 148–150, 164
Cathedral Rock, S.Africa **221**, 222
Catherine L. Ruwenzori 184, 229
cauldron subsidence 53
causses 125
cave deposits, *see* calcite formations
cave, Limestone, *see* limestone cave
cave, sea 199, **200**, 204, 207, 217, 219–220, 224
Chad Basin 11, 146, 162
Chad L. 152, 229, 233, **234**, 241
Chaillu Massif, Gabon 59–60
chalk 125
Chambe Plateau, Malawi **57**, 58, 65
Charlotte Falls, Sierra Leone 100–101
Chech, Erg, Sahara 152, 160
Chemchane Sebkha, Mauritania 163, 166, **167**, 168
chemical erosion 80–1, 196–7
chemical weathering 61–8, 81, 120, 124–8, 131–2, 135, 142, 158, 162, 168, 170, 192, 196, 198, 205
chott 233
Chott ech Chergui, Algeria 29, 233
Chotts Plateau, Atlas Mts. 29, 30
Chunya Scarp, Tanzania 21, **22**
cinder cone 37, 39–41, **43**, 44, 234
cirque, erosion, *see* erosion cirque
cirque, glacial 177–9, 183–5, 187–8, 192, 229, 232
Cirque Poumpi, Congo 76, **77**
clay 3, 4, 7, 26, 29, 56, 62, 69, 72, 74, 107, 117, 122, 158, 176, 204
cliffs, marine **7**, 27, 69, 74, **75**, 110, 194, **197**, 198–200, 202, **203**, 204–5, **206**, 207, 217, 219–220, 222, 224–6
climatic change 10–13, 138, 168–171
clints 125
coastal landforms 197–225
coastal classification 226
coastal dunes 202, 204, 213, **215**, 216, 220, 225
coastal lapiés 224
coastal plain 217, 219, 226
collapse doline 126, 128, 130–2, 238, 239

columnar jointing **7**, **45**, 114, 202
Comoro Is. 16, 36, 38–9, 224–5
composite cone 37, 41, 44–6, **47**, **188**, **236**
compressional force 15, 27
cone karst 127, 132–3
congelifraction, *see* frost weathering
congeliturbation 190
Congo Basin 10–11, 90, 112, 116, 137, 241
Congo R. 83, 90–1, 94–5, 112–13, 115, 237, 241
consequent (normal) faultline scarp 25–6
continental drift 10, 11, 92, 171–2, 226
continental rocks 2, 5, 8–11, 19, 34
continental shelf 2, 8, 92, 95, 196, 217, 219
coral 10, 196, 222
coral coasts 194, 196, 222–3, **224**, 225–7
corestones 143, 146
corrie, *see* cirque, glacial
crag and tail 175, 188
crater 35, 37, 38, 41, **42**, 46, 73
crater lake 44, 47, 105, 233–4, **235**
creep 70–1, **74**
crescent dune, *see* barchan
crevasse 174, 176
cross profile, river, *see* rivers
Cross, R., Nigeria 95, 122–3, 124
crust, earth's 1, 2, 6, 8, 9, 11, 14–15, 19, 27, 31–2, 34–5, 92
cuesta 68–9, 75, 101, 117–18
cuirass, *see* duricrust
cumulo dome 37, 41, **43**, 44, **111**
Cunene R., Angola 102–3, 153
cusp, beach 201, 210, **211**, 220
cuspate foreland 211, 214–15
cut-off 84, 91, 98
cycle of erosion, *see* geomorphic cycle

dambo 83
Danakil Alps, Ethiopia 21
Danakil Lowlands or Depression 21–3, 74
debris cone, *see* alluvial cone
debris fan, *see* alluvial fan
deep (subsurface) weathering 65, 142–4
deferred tributary 87, 89, 91
deflation 156, 158, 168, 238
deflation hollow, *see* blowout
degradation 1
delta 8, 91–5, 176, 207, 226, 229, 237
dendritic drainage 104
Densu R., Ghana 85, 91, 103
denudation 1, 6, 8, 11–13, 31, 52–3, 69, 138, 165; (*see also* weathering and erosion)
deposition
    ice 1, 2, 174–180, 228, 232
    marine 1, 2, 91, 194, 196, 200–1, 207–9
    rivers and running water 1, 2, 79, 81–96, 106, 108, 156, 160, 196, 212, 216, 228, 237
    wind 1, 2, 155, 196
deserts 152–170
    characteristics 152–4
    denudation 155–6
    landforms 138, 156–170
    weathering 62, **63**, 67, **155**, 156
desert plain 163, 164, 168
desert varnish 67
Devils Rock, The, Niger **156**
diapir 32
differential erosion 15, 25, 31, 38, 52–4, 59, 114, 117, 120–3, 157, 164, 168, 170, 175, 195, 198–9
differential weathering 64–5, 67, 121, 127, 132, 142–3, 156, 164, 168, 170
discordant drainage 105–9
dissected plateau 50, **52**
dissected volcano 38, 46–8, **50**, 224

divide migration, *see* watershed retreat
Djado Plateau, Niger 163, 165
Djourab, Erg du, Chad 152, **162**, 163
dolerite rocks 2, 54–6, 166, 202–3
doline 125–6, 128–9, 130–3, 135, 239
dolomite rocks and scenery 3, 62, 123–5, 128–9, 133, 135, 204–5, 238
dome 31–2, 52, 104–5, 108
domed inselberg 65, 140–1, 146–9
donga 72
Douentza Massif, Mali 119–120
downwash 69–70, 72, 76, 82, 127, 165, 197
draa 159–160
drainage diversion 23–4, 101–4, 175, 234
drainage patterns 23, 104–9
Drakensberg Mts. 50–1, **52**, 67, **82**, 138, 172
Drakensberg Volcanics (lavas) 4, 5, 10, 50, 51, 54
drift, *see* glacial drift
drowned coast, *see* submerged
drumlin 176
dry valley 101, 125–9, 156
dunes, desert, *see* also coastal 154–5, 159–163;
duricrust 68–9
Dwyka Tillite 5, 26, 171, 189
dyke 35, 38, 53–4, 56, 121

earth hummock (mound) 191–2
earth movements 1, 2, 4, 6, 10–11, 13–16, 29, 106, 216–17, 228, 233
earth pillar 72, 74, 76, 78
earthquake 10, 32–3, 38, 70
earthquakes in Africa 32–3
East African Rift Valley System 16–24, 32
Eastern Rift **14**, 16–19, 36, 39, 46–8, 233
Eboga caldera, Cameroon 46–7
edeyin 154
Edward L., *see* Amin L.
El Asnam earthquake 33
elbow of capture 101, 104
Elgeyo Scarp, Kenya 17–19, 21
Elgon, Mt. 24, 105, 172
El Koub ou Djaouf, Algeria 56
emerged coast 217, 219–220, 226
Enchanted Lakes, Mt Kenya 181, **228**, 229
endogenetic processes 1
End-Tertiary Surface 138, **140**, 144–6, 148–9, 163, 165
erg 154–5, 159, 160
erosion 1, 6, 12, 14–15, 21, 23, 25–9, 38, 46, 48, 50–2, 54, 56, 58–9, 61, 106, 139, 141–3, 145–6, 155, 168, 189, 224, 226, 228, 240
   ice 61, 69, 174–180, 183, 217, 229
   marine 61, 91, 93–4, 110, 194–202, 204–7, 209–212, 215, 217
   rivers and running water 11, 12, 21, 26, 28, 61, 69, 79–84, 86–7, 93, 96, 101–3, 105, 106, 110–17, 127, 156, 163, 168, 170, 179
   wind 61, 155–8, 168, 170, 238
erosion cirque 72, 76, **77**, **78**
erosion pillar, *see* earth pillar
erosion scarp 15, 20, 58, **118**, 119, 122, **123**, 138, **139**, 141–2, 145–6, 152, 166, 168
erosion surface, *see* planation surface
erratic 175, 182, 189
Erta Ale, Ethiopia 36, 38, 41
escarpment, *see* erosion, fault & faultline scarps
esker 176–7
estuarine delta 92
estuary 92, 94–5, 196, 197, 210, 213, 216–17
Etang d'Ambala, Madagascar 237
Ethiopian Highlands 16, 23, 52, 67, 172, 192
Ethiopian Rift 16–18, 48, 234
Ethiopian Scarp 16, 23
Etosha Pan 102–3, 153–4, 158

Europa Is. 225, **227**
eustatic change 13, 216–17
exfoliation 65, **66**, 120, 143, 146, 148, **155**
exhumed or resurrected relief 27, 106, 189
exogenetic processes 1
explosion crater **36**, 37, 39–40, 43–4, 233–4
extrusive landforms 34–52

faults & faulting 1, 6, 8–11, 14–27, 29–32, 34, 38, 69, 70, 83, 101–2, 108, 126, 128–9, 132, 135, 158, 199, 233
faulting & drainage 16, 18–19, 23–5, 27–8, 83, 101–2, 104–5, 108, 110, 112
faulting & landforms 14–23, 25–7, 30, 132, 135
fault-guided valley 18, 19, 25, 27–8, 105
faultline scarp 15, **25**, 26–7
fault scarp **14**, 15–19, 21, **22**, 23–4, 27, 33, 110, **131**, 132, 135, 233
faulting & coasts 25, 27
ferricrete, *see* lateritic duricrust
fiord 179–180, 196, 217, 226
firn 172
Fish River Canyon, Namibia **113**, 114, 153
fissure 34–5, 38, 51, 53
fissure eruption 35, 37, 40–1, 50
floodplain 81, 83–7, **88**, 89, 90–2, 96, 98, 103, 109, 229, 237
floods, *see* rivers
fluvioglacial outwash deposits 174, 176–7, 179, 182–3, 185, 188, 189, 229
folding 1, 4, 6, 9–11, 13, 20, 27–31, 104–6, 120, 122, 126, 128, 202, 204
folding & landscape 28–30
fold mountains 4, 10, 28–31, 150
foliation of rocks 5, 65
Fonjay Massif, Madagascar 56
fossils 2, 3, 10
Fouta Djallon 54–5, 68
freeze-thaw 62, 65, 67, 70, 190–2
fringing reef 223–5
frost weathering 62, **65**, 67, 68, 190–2
fulls & swales, *see* ridges & runnels
fumarole 36, 46
Funduzi L., S.Africa 236–7

gabbro rocks & scenery 2, 54, 56–7, 117
Galma R., Nigeria 145, 237
Gambaga Scarp, Ghana 119
Gambia R. 95, 213
gara 164
gendarme 178, 181, 183, 185
geo 199–200
geological structure of Africa 10, 11
geological time scale 4, 5
geomorphic cycle 11–13, 118, 138–9, 141, 143
geomorphic processes 1, 6
Gilf Kebir Plateau, Egypt 120, 152–3, 157, 160, 163, 165–6, 168
glacial deposits 10, 172, 174–185
glacial drift 171, 174, 175, 178–9, 189, 232, 233
glacial erosion & deposition, *see* ice
glacial periods 13, 171, 175, 216–17
glacial rock step 178–182, 185, 187
glacial trough 21, 110, 178–183, 184–8, 228, 229, 233
glaciated highland landforms 177–188
glaciated lowland landforms 175–7
glacier 69, 171–5, 177–9, 182–5, 187, 197, 217
glaciers in Africa 4, 172–4
gneiss 3, 4, 21, 54, 58, 140, 146, 149
Gondwanaland 10, 20, 138, 171, 172
Gondwana Surface 138, 144–5
Goree Is., Senegal 202, 226
gorge 24, 26, 50–1, 83, 107–8, **109**, 110–14, 121, 133, **134**, 234, 241
Gorges Valley, Mt. Kenya 180, **181**, 182–3, 188, 192

Gouffre de Lukwila, Zaire 76, **78**
Grand Dhar, Mauritania 166–7
Grand Erg Occidental 152
Grand Erg Oriental 152, 154, 160
granite rocks & scenery 2, 4, 5, 8, 10, 21, 27, 52, 54, 56–60, 64, 104, 108, 117, 120, 140, 146, 148–9, 163, 189, 226
granular disintegration 62
Great Berg R., S. Africa 102–3
Great Dyke, Rhodesia (Zimbabwe) 57
Great Eastern Erg, *see* Grand Erg Oriental
Great Escarpment 29, 31, 50–1, 82, 138, 139, 155
Great Sand Sea, Egypt 152, 157, 160, 165
Great Ruaha R., Tanzania 108–9, 113
Great Western Erg, *see* Grand Erg Occidental
grike 125
Gris L., Ruwenzori 184, 187, 192, 232
ground ice 190–2
ground water 7, 8, 62, 68, 124, 127, 135, 192
Guelb er Richat, Mauritania 123–4
gully & gullying 40, **66**, 72, 75, **76**, 95, 122, 236

hamada 153
Hamada du Guir, Sahara 152–3
Hamada el Homra, Sahara 152–3
Hamra Wadi, Gilf Kebir, Egypt 165, **166**
Hanging Tarn, Mt. Kenya 180–1, 188, 229
hanging valley 110, 178–9
Haruj el Aswad, Libya 39, 50
headland 195–6, **197**, 198, 202, 204–5, 207, 217, 220
headward erosion 72, 81, 101
Hercynian Mountain Building 28
Hex River Mts., S. Africa 31, 103
High Atlas Mts. 29, **30**, 172
High Plains (High Plateau), Nigeria 58, 138, 144–5, 146
Hobley Valley, Mt. Kenya 180–1, 183, 185, 188, **288**
Hohnel Cirque, Mt. Kenya 184–5
Hohnel L., Mt. Kenya 184, 232
hogback 117, 118
Hoggar Mts. 39, 43–4, 68, 152, **153**, 156, 163, 168–9, 171
Hombori Mts., Mali **119**, 120, 150
hooked spit, *see* recurved spit
horn 178, 181, 183, 185, 187–8
horst 15, 18
Hoyo Caves, Zaire **134**, 135
hum 126, 129
humid tropics 61, 63–5, 72, 81, 116, 118, 124–5, 127, 142, 196, 198, 201, 204
Hut Tarn, Mt. Kenya **183**, 185, **231**, 232
hydration 62, 65
hydraulic action
    rivers 80–1, 110
    waves 195, 199
hydrolaccolith, *see* ice mound
hydrolysis 61–2
hypabyssal rocks 2

ice ages in Africa 4, 171, 189–190, 192
ice-contact features 176–7
ice deposition 1, 2, 174–5, 176–7, 178–9, 180, 228, 232
ice erosion 61, 69, 174–180, 183, 217, 229
ice-eroded plain 175, 189, 229
ice formation 172
ice mound 191–2
ice movement 174
ice sheet 13, 171–2, 174–7, 189, 232
ice wedge 191
igneous rocks 2, 6, 34–60, 61, 63, 65
Iharen, Hoggar Mts. 44, **45**
Imo R., Nigeria 27, 122–3
incised meander 26, 98–9
inconsequent drainage, *see* discordant drainage
ingrown meander 98, 99

inland delta 94, 119, 241
inland drainage basin 164, 165–8
inselberg 59–60, 65, 117, 137, 139–50, 163, 164, 168
interglacial period 13, 171, 175, 179, 217
interlocking spur 82, **83**
intermediate lava 34, 37, 44
intermontane basin 29
intermontane plateau 29
intrenched meander 98, 99
intrusive landforms 34, 52–60
inverted relief 29, 121
Isalo Scarp, Madagascar **25**
islands 8, 10, 38–9, 48, 200, 202, 211, 214, 217, 223–6
isostasy 8
isostatic adjustments 8, 13, 92, 217, 223
Itasy L. Madagascar **43**, 44, 235, 240
Itasy Massif, Madagascar 41, **43**, 44

Jebel Marra, Sudan 3, 39, 152
jointing of rocks 6–8, 50–1, 62, 65, 67, 70, 81, 83, 105, 112, 117, 120, 124–5, 126–7, 132, 135, 142–3, 146–9, 168, 170, 174–5, 177, 179, 195, 198–200
Jos Plateau 39–40, 58, 138, 144, **145**, 146, 147, 240

Kaap Plateau, S. Africa 135, 153
Kabarega Falls (Murchison Falls), Uganda 24, 112
Kaduna R., Nigeria 81
Kafu R., Uganda 24
Kafue R., Zambia **83**
Kagera R., Uganda 24, 83, 233, 236
Kainji L., Nigeria 109, 239
Kalahari Desert 13, 69, 152–4, 158–9, 168–9
Kalambo Falls, Zambia **23**, 24, 110
Kamasia Ridge, Kenya 17–19
kame 176–7
kame terrace 177
Kamusoso Valley, Ruwenzori 184, 187–8, 229
Karas Mts., Namibia 21, 153
Kariba L. 16, 239
Karum L., Ethiopia 16, 32
karst 117, 123–135, 238–9
Kartala, Ngazija (Grand Comoro) 36, 38, 41
Katwe L., Uganda 39, 40
Kedong Scarp, Kenya 17
kegelkarst, *see* conekarst
Kenya, Mt. 17, 48, **65**, 67, 172, **173**, 174, 180–5, 188, 190, 192, 229, 231, 232
Kerio R., Kenya 17, 18, 19
Kerrata Gorge, Algeria 109
kettlehole 176–7, 187, 232–3
Kibo, Kilimanjaro 17, **35**, 45–6, **47**, 174, **188**
Kilimanjaro 17, 35–6, 45–6, **47**, 49, 150, 172, 174, **188**, 192
Kilombero R., Tanzania 90–1, 96
Kinkon Falls, Guinea 54, **55**, 110
Kissenga Plateau, Congo 132–3
Kitsimbanyi, Zaire 40, 42
Kitia L., Madagascar 43, 44
Kivu L., Zaire **6**, 18, 24, 41–2, 83, 235
Klein Spitzkopje, Namibia 149, **163**
Klein Berg R., S. Africa 27
knickpoint 98–9, 101, 111, 116
Ksouatene Polje, Morocco 128–9
Kudaru Hills, Nigeria 58
Kufena Hill, Nigeria **146**
Kuiseb Canyon, Namibia 153, **161**, 162
Kwahu Scarp, Ghana 112, 119
Kyoga L., Uganda 24

laccolith 53, 56
lagoon 91, 93, 194, 201–2, 204, 207, **208**, 209–16, 220, 223–5, 237–8
lahar 73, 74

Laikipia Scarp, Kenya 18, 33
lakes & lake basins 16, 23, 39, 43, 73, 82, 85, 87, 90–2, 94–5,
    106, 110, 114, 126, 128, 149, 158, 166–7, 175–7, 179–185, 215,
    228–241
landslides 33, 70–2, **73**, 110, 198, 205, 236
Langue de Bargarie, Senegal 211–12
Lanzarote Is. 36, 41
lapiaz 125
lapiés 125, 128
lateral erosion 81, 83, 96, 97, 178
lateral moraine ridge 176, 179–185, 188, **231**, 233
lateritic duricrust 66, 68–9, 72, 118, 122, 145, 202
lava 34–5, 38, 43, 49–51, 72, 112
lava flow **6**, 17, 21, 38–43, 46, **49**, 233–5
lava plateau 49–51, **52**
lava sink 41
Lavaboro Sink, Madagascar 129–130
lavaka 76, 93
levée 85, 87, 89–92, 94, 237
Lewis Glacier, Mt. Kenya **173**, 174, 182–3, **184**, 185, 192
Likabula R., Malawi 54, 58
Likaiyu, Kenya 36, 39–40
Lily Falls, Madagascar 44, 110, **111**, 235
limestone 3, 4, 29, 62, 81, 117, 123–135, 168, 196–7, 205–6, 222,
    224
limestone landorms 125–135, 238–9
limestone cavern, cave 125, 128–133, **134**, 135, 238
limestone gorge 113–14, 127, 133, **134**, 135
limestone pavement **125**
limestone solution 124–8, 131–2, 135, 196, 205, 224, 238–9
Linta R., Madagascar 129–130
Livingstone, Mts. 17, 21
Lobé Falls, Cameroon **110**
Lomati R., S.Africa 121–2
longitudinal dune, see seif dune
Longonot, Kenya 17, 36, 47, **49**
long profile, river, see rivers
longshore drift 195, 208–212, 219, 222, 238
lopolith 53, 56–7
Loza R., Madagascar 218
Luangwa R., Zambia 16, 17, 27
Lumley Beach, Sierra Leone 214
Lupata Gorge, Mozambique 7, 114

Mackinder Valley, Mt. Kenya 180, 192
Madeleine Bay, Senegal 202, **203**
Madeleine Is., Senegal 202, 226
Maghra L., Egypt 168, 238
magma 34–5, 37–8, 52–3
Mahafaly Plateau, Madagascar 129–131
Mahoma L., Ruwenzori 187, 233
Makarikari Depression, Botswana 153, 168–9, 240
Makhonjwa Range, S. Africa/Swaziland **121**, 122
Makteir, Erg; Mauritania 163, 166–7
Malawi L. 16–17, 19, 23, 24, 74, **232**, 233, 240
Malawi Rift 16, 17
Maletsunyane R., Lesotho 83
Manambolo R., Madagascar 114, 133, **134**, 135, 237
Mananjeba R., Madagascar 131, 132
Mandrare R., Madagascar 215–16
Mangoky R., Madagascar 26, **86**, 92
mangrove 91–3, 194, 196, 212–14, 216, 218, 225–6
man-made influence 69, 72, 87, 94, 194, 197, 209–10, 212, 239–40
mantle 1, 8, 9, 10, 19, 34
Margherita Peak, Ruwenzori 21, 184, 187
marine deposition & erosion, see waves
massive rocks 7, 69–71, 104, 120
mass-wasting 1, 61, 69–76, 82–3, 127, 141, 165, 168, 197–8, 205,
    236
Matapo Hills, Rhodesia (Zimbabwe) **59**, 60
Matupi Cave, Zaire **134**, 135
'maturity' 12

Mau Escarpment, Kenya 17, 19
Mawenzi, Kilimanjaro 17, 45–6, **47**, 49, 188
Mayotte Is 224–6
Mbashe R., see Bashee
meander 81–3, **84**, 85, **86**, 87, 89, 91, 96, 98
meander migration 84–5, 97
meander scar 85–7, **88**, 89
meander terrace 97–8
mechanical weathering, see physical
Medjerda R., Tunisia 93, 212–13, 226
meltwater deposits 176–7
Menengai, Kenya 17, 47
Mercalli Scale 32–3
Meru, Tanzania 17, 36, 46, 74
mesa 118–120, 145
metamorphic rocks 3, 4, 6, 63–5
meteorite crater lake 239
Mgahinga, Uganda 42, 44, **236**
Mgeni R., see Umgeni
Michaelson L., Mt. Kenya 180–2, 192, 229
Middle Atlas Mts. 87, 128–9, 239
misfit stream 101, 178
Mobutu L. 16–21, 24, 39, 112, 233, 240
Moka-Long Range, Mauritius 48, **50**
mokgacha 168
Molopo R., S.Africa 69, 153, 158, 169
Moore Glacier, Ruwenzori 174, 185
moraine ridge 69, 74, 171, **173**
Mosi oa Tunya, see Victoria Falls
mountains & mountain regions 9, 11, 28–9, 35, 46, 50–2, 62,
    65, 67–8, 70, 74, 150, 152–3, 171–2, 178, 190–2, 196
mountain building periods 4, 28–9
mountains, classification 150
Mubuku Valley, Ruwenzori 21, 184–5, 187, 233
Muchinga Escarpment, Zambia 26–7
mud flow 70–3, 76, 164
Mufumbiro Volcanoes, see Virunga
Muhavura, Uganda 40, 42, 44, 234, **236**
Murchison Cataracts, Malawi 24
Mutanda L., Uganda 40, 42, 235, **236**
Mutito Scarp, Kenya 27
Mwachi R., Kenya 99

Nabuyatom ash cone, Kenya 40
Naivasha L., Kenya 17, 18, 240–1
Namib Desert 139, 149, 152–6, 160, **161**
Namuli Mts., Mozambique 142, 150
Nasser L., Egypt 93, 157, 163, 239
natural arch, see arch
nebka 215–16
neck, volcanic, see volcanic neck
negative movement of base level 12, 96, 116, 216–17
Nelion, Mt. Kenya 48, **65**, **173**, **183**, 184–5
névé, see firn
Ngaila R., Kenya **84**, 91, 98
Ngaliema Mt., see Mt. Stanley
Ngorongoro, Tanzania 17, 48
Ngozi L., Tanzania 233–4
Niger R. 6, 79, 89, 91–2, 98, 109, 118–120, 122–3, **137**, 138, 144,
    146, 168, 216
Nile R. 27, 83, 90, 92, **93**, **107**, 108, 113, 125, 152–3, 157
Nimba Mt., Guinea 121
Nithi R., Kenya 180–2
nivation 67, 192
nivation hollow 177, 192
Njorowa Gorge, Kenya 7, 241
Nkontompo Stack, Ghana 205, **206**, 207
Noir L., Ruwenzori 184, 187, 229, **230**
Nokoué Lagoon, Benin 207–8, 238
normal fault 15, 19, 20
Nsaki R., Ghana 103–4
nunatak 174

Nyabihoko L., Uganda 24, 236
Nyamlagira, Zaire **6**, 24, 36, 38, 40–1, **42**
Nyando R., Kenya 91, 98
Nyasa L., *see* Malawi L.
Nyika Plateau, Malawi 104, 138
Nyiragongo, Zaire 36, 38, 46
Nyungu L., Uganda **36**, 40, 234

obsequent (reversed) faultline scarp 25–6
ocean ridge 8–9
ocean trench 8–10, 38
oceanic rocks 1, 8–10
offloading, *see* pressure release
Ogooué R., Gabon 60, 237
Ogun Hills, Nigeria 149
Okavango Delta, Botswana 69, 94, **95**, 102, 153–4
'old age' 12
Oldoinyo Lengai, Tanzania 17, 20, 36, 38, 44
Ojikoto L., Namibia 238, **239**
Omo R. Ethiopia 16, 94
Orange R. 51, 82, 90, 114, 138, **140**, 153–5
Orugu R., Sierra Leone 100–101
Oti R., Ghana 98
outlier 120, 122, 164
outwash plain 177
overflow gorge 113, 114, 241
Owen Fracture 9, 27
oxbow lake 86–7, 89, 91, 237, **238**
oxidation 61–2, 64, 162
Oyster Bay, Tanzania 224

paired terrace 97
paleomagnetic dating 10
pan, *see* blowout
Pangaea 10, 11
parabolic dune 215–16
parasitic cone 35, 37, 41, 44–6, **49**, 188
past climates 13, 68, 124, 139, 152, 155–6, 163–4, 168–170
patterned ground 191–2
pediment 140–2, 156, **163**, 164, 166
pediplain 142
pediplanation 141–3
pedimentation 141
Penck Glacier, Kilimanjaro 174
peneplain 141
peneplanation 141
periglaciation 171–2, 175, 190–2
periglacial landforms 67, 175, 190–2
peripediment 166
permafrost 190–2
phreatic zone 7
physical weathering 62–3, 65–8, 190–2
Pico de Teide, Canary Is. 46
Pieter Both, Mauritius **50**
pingo 192
piprake 191
Piton de la Fournaise, Réunion **35**, 36, 38, 41, 48
Piton des Neiges, Réunion 48
plains 11, 12, 83, 107, 111, 116, 118, 137–49, 154, 163, 177, 217, 219
planation surface 138–49, 163, 165
planezes 38, 48
plate tectonics, *see* tectonic plates
playa, *see* sebkha
plateau 29, 49–52, 58, 68, 118–20, 129–32, 152, 164–5, 169
Pleistocene Ice Age 13, 63, 171, 172, 189, 192, 197, 208–9, 217
Plettenberg Bay, S.Africa 31, 100, 220, **221**, 222
plug dome 37, 43, 44, **45**, **153**
plutonic rocks 2, 54, 58
pluvial period 13, 164, 168
point-bar deposits 85, **86**
Point John, Mt. Kenya **173**, 183, 185, **231**

Pointe à Larée, Madagascar 214–15
polje 126, 128–9, 239
polycyclic landscape 139
polygonal jointing 7; (*see* also columnar jointing)
ponor, *see* sink
positive movement of base level 12, 96, 216–17
pothole 81
pressure release jointing 6, 65, 174, 177
pyramidal peak, *see* horn
pyroclasts 34–5, 37, 44, 50, 75, 234

Qattara Depression, Egypt 152, 157, 168, 238
qoz 169
Quaternary Surface 138, 144, 146
quartzite rocks and scenery 4, 56, 65, 108, 117, 120–23, 128, 189, 204

radial drainage 104–5,
raised beach 4, **12**, 13, 194, 217–19, **220**, 226, 240
rapids 24, 69, 81–2, 99, 100, 103, 108–113, 116
Ras Hafun, Somalia 16, 212, 214
recessional moraine 176, 185
rectangular drainage 25, 104–5
recurved spit 93, 211–13
Red Sea 11, 152, 201, 215, 224
reg 154–6
rejuvenation 12–13, 81, 83, 96–101, 111, 114, 142
rejuvenation gorge 98, **100**, 114
rejuvenation terrace, *see* river terrace
residual 108, 125, 126, 138, 140–1, 163–5
residual mountain 150
residual volcanic landforms, *see* vulcanicity
resurgence 128–9, 132
resurrected relief, *see* exhumed
reversed drainage 23–4, 106, 233
reverse fault 15
Rhumel Gorge, Algeria 109, 114
rhourd 160
ria 217–18, 226
Ribi R., Sierra Leone 56, 217, **218**
Richter Scale 32–3
ridges and runnels (beach) 201, 210
Rif Atlas Mts. 29–30, 73, 87
rift faulting 4, 10, 17, 23
rift valley 11, **14**, 15–24, 27, 33, 38, 233
rills 72
ring complex 53, **57**, 58–9, 107–8, 145
ring crater, *see* explosion crater
ring dyke 53, 58–9, 108
rivers 79–116
    capture 25, 96, 101–4
    channel 79–81, 83–7, 89, 92, 110–11, 114, 141–2, 179
    cliff 83–5
    competence 79, 82
    deposition 79, 81–96, 106, 108, 196, 212, 228, 237
    discharge 13, 79, 81, 83, 85, 86, 89, 94, 96
    energy 79–81, 83, 94–7, 101
    erosion 11–12, 21, 26, 28, 69, 79–84, 86–7, 93, 96, 101–3, 105–6, 110–17, 127, 179
    floods 79, 81, 83, 85–7, 89, 92, 162, 237
    load 79–83, 86, 92–5, 97, 116
    lower course 79, 81, 83–91, 96, 110, 114, 116, 237
    profiles 82–3, 98, 100, 112, 114–16
    terrace 4, 96–8
    transport 79–80
    upper course 82–3, 110, 113, 114
roche moutonnée 175, 178, 185, **187**, 188, 189
rocks 1–8
    dip 2, 28, 69, 104, 117–18, 122, 199
    hardness 63–5, 83, 104–5, 108, 110–11, 113–14, 116–17, 124
    impermeability 7, 8, 117, 119
    permeability 7, 8, 117, 124, 199, 234

perviousness 7
porosity 7
structure 8, 11, 14–32, 65, 69–71, 101, 104–7, 110–12, 114, 117–124, 126, 128–9, 202, 204
rock basin 149, 173, 175, 177–181, 184, 187–9, 228–9, 232
rockfall 70–1, 74, **75**, 175
rock glacier 70, 75
rock pedestal **156**, 164
rock salt 3, 32, 72, 74
rock slide 70–2, 75
rock slump, *see* slump
rock tower **164**
Romanche Fracture 9
rotational slip 178
rotational slumping 71–3
Ruacana Falls, Angola 103
Rudolf L., *see* Turkana L.
Rungwe Mt., Tanzania 43, 74, 233
ruware 149
Ruwenzori 16–17, **20**, 23, 67, **80**, 96, 150, 172–4, 184–8, 190, 182, 229–230, 232–3, 236

Sabaloka Gorge, Sudan **107**, 108, 113
Sabaloka Hills, Sudan 59, 107, 108
Sabi R., Rhodesia (Zimbabwe) 91
Sabinio, Uganda 42, 74, **236**
Saddle, The, Kilimanjaro 45, 46, **188**, 192
Saddleback Range, S.Africa 121–2
Sahara 3, 13, 67–8, 139, 152–7, 160, 162–171, 192
Sakumo Lagoon, Ghana **208**
salt crystallisation weathering 62, 67, 170
salt dome 32
saltation 156
sand cay 226
sand dunes, *see* dunes
sand formations 159–163
sand ripples 159
sandstone rocks and scenery 3–7, 25–7, 29, 54, 64, 68, 73–6, 99, 102, 106, 114, 117–120, 122–4, 128, 132–3, 157, 163–6, 169–170, 197, 198, 204–5, 236
San Juan volcano, Canary Is. 38
Santa R., Guinea 27
savanna lands 66, 68, 70, 72, 81, 82, 138, 142
scarp, *see* erosion, fault & faultline scarp
scarplands 104, 118, 122–3
scarp retreat 141–2, 145, 165
schist 4, 5, 21, 54, 60, 121, 123, 204
scoria cone, *see* ash and cinder cone
scree, *see* talus
sea floor spreading 8
sea level change 12, 13, 92, 96, 110, 194, 196, 216–220, 222, 226
sea mount 223
sebkha 166–7
Sebou R., Morocco 87–9
sedimentary rocks 2, 3, 6, 62, 64–5
seif dune 159–160, **161**, 162, 168
Selima Sand Sheet, Egypt 152, 157, 162, 165
Semien Mts., Ethiopia 172
Semliki R., Zaire 20, 21, 96
Senegal R. 211–12
serir 154
Serra da Chela, Angola 103, 121
Seychelles 10, 225–6
Shala L., Ethiopa 16, 48, 234
shale 3–5, 69, 117, 121, 158, 204–5
Share Cuesta, Nigeria 68–9
sheetflooding, sheetwash 68, 72, 75, 83, 141, 156
shield volcano, *see* basalt dome
shingle beach 196, 201
Shira, Kilimanjaro 17, 46
Shira Ridge, Kilimanjaro 45, 46
Shire R., Malawi 24, 54, 146

Shire Valley, Malawi 16, 17, 19
sial 2, 8, 19
Sierra Leone Peninsula 56, 95, 219
Sierra Leone R. 56, 217
silcrete 68–9
sill 53–4, **55**, 56
sima 8
sink or ponor 126, 128–132, 135
Siwa Depression, Oasis, Egypt 157, 160, 168
slip-off slope 83–5, 98
slopes 7, 21, 41, 43, 51, 69–72, 104–5, 117, 120, 129, 140–2, 153, 164–6, 176, 199, 204–5
   movement on slopes 68–78, 168
slump & slumping 70–2, 74–6, 81, 196, 199, 205
Sof Omar Caves, Ethiopia 135
soil creep, *see* creep
solifluction 71–2, 74, 190–2
Souffleur, Le, Réunion 207
spalling 65
Speke Glacier, Ruwenzori **20**, 174
Speke L., Ruwenzori 184, 188, 229
Speke, Mt., Ruwenzori 21, 174, 184, 229
spheroidal weathering 65
spit 93, **161**, 208, 210–11, **212**, 213, 216, 219, 226, 240
St Mary's Is., Gambia 212, **213**
stack 199–200, 202, 204–5, 207, 220
stage of erosion 12, 118
stalagmites & stalactites 128, 132, **134**, 135
Stanley, Mt., Ruwenzori **20**, 21, 67, **80**, 174, 184, 187
stepped faulting 17, **18**
stone garland 191
stone polygon 191–2
stone stripe 191–2
Storms R., S.Africa **100**
stratified drift 176–7
strato-volcano, *see* composite cone
stream diversion, *see* drainage diversion
striated rock 171, 185, 187–9, **190**
structural basin, depression 11, 19, 90, 116, 119, 137, 139, 155, 158
structural bench 113–14
structural swell 11, 19, 139
structure of continents & ocean basins 8
subduction zone 8
submarine canyon 95
submerged coast 96, 179, 194, 208, 217–220, 225, 226, 238
subsidiary cone & vent 36–7, 41–2, 46
subterranean stream, *see* underground drainage
Suguta Valley, Kenya 39, 41, 240
superimposed drainage 106–8, 113, 114
Suswa, Kenya 17, 47
Swakop Canyon, Namibia 161, 162
swallow hole 126
Swartberg, S.Africa 31, 135
swell, ocean, *see* waves
swell, structural, *see* structural swell
synclinal upland 29, 31, 105
syncline 27, 29, 31, 105, 121, 204

Table Mt., Cape Town 31
Table Mt., Natal 26
taffoni 67
talus 56, 62, 70–1, 146, 178
talus creep 70–1, 74
talus slope 26, 49, 62, **63**, 67, **118**
Tana R., Kenya 90–1, 237, **238**
Tanezrouft, Sahara 152–3, **154**, 163
Tanganyika L. 16–17, 23–4, 33, 171, 229, 233
Taoudene Basin 11, 119
tarn 180
Tassili N'Ajjer, Sahara 152, 168, **169**, 170
tear faulting 23, 27

tectonic plates 8–11, 19, 27, 32, 34, 38
Teleki volcano, Kenya 36, 40
Teleki Cirque, Mt. Kenya **183**, 184–5
Teleki Tarn, Mt. Kenya 183, 185, 229
Teleki Valley, Mt. Kenya 173, 180–1, **182**, 183, 185, 192, 232
Tell Atlas Mts. 29, 30, 96
Ténéré, Sahara 152–3, 163
tensional force 15, 19–20
terminal moraine ridge 176–7, 179, 182–5, 187–8, 228–9, 232
tholoid 43
Three Sisters, S.Africa 54, **55**
thurfur, *see* earth hummock
Thyolo Scarp, Malawi 54
Tibesti Mts. 39, 46–9, 67, 152–3, 162–4
tidal currents 91–2, 194, 196, 210
Tikjda Caves, Algeria 135
till 174–7, 179, 187–9
till plain 175–6
tillite 3, 171, 189
tilt block landscape 16, 21, **22**
tilting of rocks 2, 16, 21, 23
Tiva R., Kenya 101–2
tombolo 202, 211, 212, 214, 220, 222
tor 141
Tororo Rock, Uganda 49
towerkarst 127, 131–33
transform fault 9
transverse dune **154**, 160, 162, 215–16
trellis drainage 104, 121
Titriva L., Madagascar 234, **235**
tropical karst 127, 131–2
Trou au Natron, Chad 47, **48**
truncated spur 178
tsunami 33
Tugela R., S.Africa 104, 139
Tulbagh Kloof, S.Africa **102**, 103
Tulbagh R., S.Africa 31, 103
Turkana L., Kenya 16–17, 39, 40, 54, 94, 233, 240
turmkarst, *see* towerkarst
Tyndall Glacier, Mt. Kenya **173**, 174, 183, 185, 229, 232
Tyndall Tarn, Mt. Kenya 183, 185, 229

Ubari, Edeyin, Sahara 152, **154**, 162
Udi Cuesta, Nigeria 75, 122, **123**, 124
Uhambingetu, Tanzania 148
Umgeni R., S.Africa 26, 54, 99
Umvukwe Range, Rhodesia (Zimbabwe) 57
underfit stream, *see* misfit
underground drainage 125–7, 129, 131–2, 135
unpaired terrace 97
unstratified drift 175–6
Urema Trough, Mozambique 16–17
Usambara Mts., Tanzania 21, 100, 150
uvala 126, 128, 130

Vaal R., S.Africa 98, 101, 108
vadose zone 7, 8, 127
valley cross profile, *see* river
valley train 179
valley within a valley 98, 100
ventifact 156
Vert L., Ruwenzori **173**, 184, 187, 192, 229
vertical erosion 81–3, 101, 106, 113

Victoria Beach, Nigeria **209**, 210
Victoria Falls 27, **28**, 111–13, 116, 153
Victoria L. 16–17, 19, 24, 83, 91, 98, 148, 228, 233, 236, 240
Virunga Volcanoes, Zaire 41, 44, 234–5, **236**
Volta Lake, Ghana 112, 138, 239, **240**
V-shaped valley **82**, 83
vulcanicity 1, 10, 11, 15, 17–18, 20, 34–60, 70, 74, 228, 233–5
   volcanoes 8, 34–49, 104, 150, 202, 223, 234–5
   types of volcano 35–8
   types of eruption 34–7, 44, 49–50
   active volcanoes in Africa 36, 41, 44, 46
   dormant volcano 36
   extinct volcano 17, 36, 40, 44, 48
   recent volcanic eruptions in Africa 38
   residual volcanic landforms 35, 38, 46–49
   vent 35
   volcanic neck 38, 44, 47–9, **51**

wadi 164–5, **166**, 169–70
Walvis Bay Spit, Namibia **161**, 212
warping 1, 4, 8, 10–11, 13, 19, 23–4, 108–9, 166, 217, 233
Wase Rock, Nigeria 49, **51**
waterfall 23, 53–5, 81–3, 100–1, 103, 110–13, 116, 139, 178, 180–2, 234
waterlayer weathering 198, 202–3
watershed retreat 101
watertable 7, 8, 62, 125–8, 131, 135, 158, 168, 238
waves 194–209
   deposition 1, 2, 91, 194, 196, 200–1, 207–9
   energy 91–2, 195–6, 198–9, 205, 209
   erosion 61, 91, 93–4, 110, 194–202, 204–7, 209–12, 215, 217
   backwash 195, 200, 209
   swash 195, 200
   refraction 195–6, 200, 211
   swell 194–5, 201, 208, 211
wave cut bench 197
wave cut platform 197, **198**
weathering 1, 12, 61–9, 81–3, 114, 117, 120–1, 141, 156, 158, 164–5, 170, 192, 197, 205
   climate & w. 63–5
   depth of w. 63, **64**, 65, 67, 142, 145
   landforms due to w. 61, 68, 124–7, 142–3, 198, 202
   plants & w. 67
   relief & w. 67
   rate & character of w. 63–8
   rock type & w. 63–5
weathering front 62–3, 143
Western Rift 16–20, **22**, 24, 36, 39, 41, 233, 235
whaleback 141, 148–9
Wilge R., S.Africa **74**, 91, 237
wind gap 101, 104
wind erosion 61, 155–8, 168, 170, 238
wind deposition 1, 2, 155, 196
wind and dune formation 159–163, 215
Wum L., Cameroon 105, 234

yardang 156, **157**
Young Fold Mountains 28, 106
Younger Granites 58, 145
'youth' 12

Zambesi R. 24, 27, **28**, 79, 83, **86**, 91, 111–12, 114, 116, 138, 239
zeugen 157
Zomba Mt., Malawi 59